DATE DUE

AP 2~ '99		
NV 27 '99		
DE~30~'0~		
NY 22 '02		

The Practical Approach Series

SERIES EDITORS

D. RICKWOOD
Department of Biology, University of Essex
Wivenhoe Park, Colchester, Essex CO4 3SQ, UK

B. D. HAMES
Department of Biochemistry and Molecular Biology
University of Leeds, Leeds LS2 9JT, UK

Affinity Chromatography
Anaerobic Microbiology
Animal Cell Culture
  (2nd Edition)
Animal Virus Pathogenesis
Antibodies I and II
Basic Cell Culture
Behavioural Neuroscience
Biochemical Toxicology
Biological Data Analysis
Biological Membranes
Biomechanics—Materials
Biomechanics—Structures and
  Systems
Biosensors
Carbohydrate Analysis
  (2nd Edition)
Cell–Cell Interactions
The Cell Cycle
Cell Growth and Division
Cellular Calcium
Cellular Interactions in
  Development
Cellular Neurobiology
Centrifugation (2nd Edition)

Clinical Immunology
Computers in Microbiology
Crystallization of Nucleic Acids
  and Proteins
Cytokines
The Cytoskeleton
Diagnostic Molecular Pathology
  I and II
Directed Mutagenesis
DNA Cloning 1:
  Core Techniques
DNA Cloning 2:
  Expression Systems
DNA Cloning 3:
  Complex Genomes
DNA Cloning 4:
  Mammalian Systems
Drosophila
Electron Microscopy in Biology
Electron Microscopy in
  Molecular Biology
Electrophysiology
Enzyme Assays
Essential Developmental
  Biology

R

DNA Cloning 2

Expression Systems

A Practical Approach

Edited by

D. M. GLOVER

*Cell Cycle Genetics Research Group, CRC Laboratories,
Department of Anatomy and Physiology, Medical Sciences Institute,
The University, Dundee DD1 4HN, UK*

and

B. D. HAMES

*Department of Biochemistry and Molecular Biology,
University of Leeds, Leeds LS2 9JT, UK*

IRL PRESS

——at——

OXFORD UNIVERSITY PRESS

Oxford New York Tokyo

Walton Street, Oxford OX2 6DP

rd   New York
and   Bangkok   Bombay
wn   Dar es Salaam   Delhi
Kong   Istanbul   Karachi
adras   Madrid   Melbourne
airobi   Paris   Singapore
Taipei   Tokyo   Toronto
and associated companies in
Berlin   Ibadan

Oxford is a trade mark of Oxford University Press

Published in the United States
by Oxford University Press Inc., New York

© Oxford University Press, 1995

A catalogue record for this book is available from the British Library

Library of Congress Cataloging-in-Publication Data
DNA cloning: a practical approach/edited by D.M. Glover and B.D.
Hames. — 2nd ed.
p. cm. — (Practical approach series; v. ‹148, 149, 163, 164›)
Includes bibliographical references and indexes.
Contents: v. 2. Expression systems.
1. Molecular cloning — Laboratory manuals. 2. Recombinant DNA —
Laboratory manuals. I. Glover, David M.   II. Hames, B. D.
III. Series.
QH442.2.D59   1995   574.87'3282 — dc20   94-26740
ISBN 0 19 963479 3 (Hbk)
ISBN 0 19 963478 5 (Pbk)

Typeset by Footnote Graphics, Warminster, Wilts
Printed in Great Britain by Information Press Ltd, Eynsham, Oxon.

Preface

It is a decade since the first editions of *DNA Cloning* were being prepared for the Practical Approach series. It is illuminating to look back at those volumes and reflect how the field has evolved over that period. We have tried to distil out of those earlier volumes such techniques that have withstood the test of time, and have asked many of the former contributors to update their chapters. We have also invited several new authors to write chapters in areas that have come to the forefront of this invaluable technology over recent years. The field is, however, far too large to cover comprehensively, and so we have had to be selective in the areas we have chosen. This have also led to each of the books being focused onto particular topics. This having been said, Volume 1 covers core techniques that are central to the cloning and analysis of DNA in most laboratories. Volume 2 on the other hand turns to the systems used for expressing cloned genes. Inevitably the descriptions of these techniques can be supplemented by reference to other cloning manuals such as *Molecular cloning* by Sambrook, Fritsch, and Maniatis (Cold Spring Harbor Laboratory, New York, 1989) as well as other books in the Practical Approach series. Volume 3 examines the analysis of complex genomes, an area in which there have been many important developments in recent years, both in the description of new vectors and in the strategic approaches to genome mapping. Again, companion volumes such as *Genome analysis* can be found in the Practical Approach series. Finally, in Volume 4 we look at DNA cloning and expression in mammalian systems; from cultured cells to the whole animal.

In the first edition of *DNA Cloning*, Volume 1 published in 1985, Richard Young and his colleagues described the use of bacteriophage lambda for cloning cDNAs in *E. coli* and their expression. The first chapter of the present volume is an extension and update of this earlier chapter and describes screening lambda expression libraries with antibody and protein probes, highlighting features that are important for successful screening. Expression using plasmid vectors in *E. coli* is covered in Chapters 2–4. The second chapter describes and compares many of the proven plasmid expression vectors for *E. coli*, offering detailed guidance on their choice and use. Many of these systems incorporate features designed to facilitate purification of the expressed protein using affinity chromatography and these are discussed at length. The approach of overproduction of proteins in *E. coli* has become a widespread tool in studying protein structure and function and in raising specific antibodies for a range of applications. Many recombinant proteins expressed at high level in *E. coli* form insoluble inclusion bodies, making it difficult to recover the protein in its native form. The re-folding of denatured proteins is also problematic. If at all possible, therefore, the aim for structural and functional studies is to produce the protein in a soluble and active form. Approaches towards achieving this goal are discussed in Chapter 3, as well as methods for solubilization of inclusion bodies for re-folding experiments. The solubilization of inclusion bodies as a means of providing large quantities of antigen to be used in raising polyclonal antibodies is covered in Chapter 2 and raising monoclonal antibodies against recombinant proteins is described in Chapter 4. Attention then turns to eukaryotic expression systems. Chapter 5 is devoted to the expression of foreign genes in yeast, and Chapter 6 describes the exciting and powerful new approach of interaction trap cloning in yeast (two-hybrid system). This allows the cloning of cDNAs encoding proteins that interact with a protein whose coding sequences are already known and so provides a ready means for studying protein–protein interaction in detail. The final chapter examines the baculovirus expression system. In recent years, this insect cell system has become increasingly popular for the over-expression of foreign post-translationally modified proteins in a eukaryotic host. Mammalian systems are an obvious omission here; as mentioned above, these now form the subject of Volume 4.

It is our hope that this book and its sister volumes will find their way onto bookshelves in laboratories and in so doing will become messy and 'dog-eared'. It has been gratifying to see how widespread the use of the first edition has become, and a similar success would be rewarding to both the editors and the authors. Finally, and most importantly, we wish to thank all of the authors for their contributions and for their patience in bringing this project to fruition.

Dundee and Leeds   David M. Glover
September 1994   B. David Hames

Contents

3. Purification of over-produced proteins from E. coli cells 59

Reinhard Grisshammer and Kiyoshi Nagai

4. Production of monoclonal antibodies against proteins expressed in E. coli 93

Steven M. Picksley and David P. Lane

5. Expression of cloned genes in yeast 123

Michael A. Romanos, Carol A. Scorer, and Jeffrey J. Clare

7. The baculovirus expression system 205

Gunvanti Patel and Nicholas C. Jones

Appendix. Addresses of suppliers 245

Index 249

Contents lists of companion volumes 254

Contributors

ROGER BRENT
Department of Molecular Biology, Massachusetts General Hospital, 50 Blossom Street, Boston, MA 02114, USA.

SEAN B. CARROLL
Howard Hughes Institute, Laboratory of Molecular Biology, University of Wisconsin-Madison, 1525 Linden Drive, Madison, WI 53706, USA.

JEFFREY J. CLARE
Department of Cell Biology, Wellcome Research Laboratories, Langley Court, South Eden Park Road, Beckenham, Kent BR3 3BS, UK.

RUSSELL L. FINLEY JR
Department of Molecular Biology, Massachusetts General Hospital, 50 Blossom Street, Boston, MA 02114, USA.

REINHARD GRISSHAMMER
Cambridge Centre for Protein Engineering/MRC Centre, Hills Road, Cambridge CB2 2QH, UK.

NICHOLAS C. JONES
Gene Regulation Laboratory, ICRF, Lincoln's Inn Fields, London WC2A 3PX, UK.

DAVID P. LANE
Department of Biochemistry, Medical Sciences Institute, University of Dundee, Dundee DD1 4HN, UK.

JAMES A. LANGELAND
Howard Hughes Institute, Laboratory of Molecular Biology, University of Wisconsin-Madison, 1525 Linden Drive, Madison, WI 53706, USA.

BENJAMIN MARGOLIS
Department of Pharmacology, Kaplan Cancer Center, New York University Medical Center, New York, NY 10016, USA.

KIYOSHI NAGAI
MRC Laboratory of Molecular Biology, Hills Road, Cambridge CB2 2QH, UK.

GUNVANTI PATEL
Gene Regulation Laboratory, ICRF, Lincoln's Inn Fields, London WC2A 3PX, UK.

Contributors

STEVEN M. PICKSLEY
Department of Biochemistry, Medical Sciences Institute, University of Dundee, Dundee DD1 4HN, UK.

MICHAEL A. ROMANOS
Department of Cell Biology, Wellcome Research Laboratories, Langley Court, South Eden Park Road, Beckenham, Kent BR3 3BS, UK.

CAROL A. SCORER
Department of Cell Biology, Wellcome Research Laboratories, Langley Court, South Eden Park Road, Beckenham, Kent BR3 3BS, UK.

JAMES B. SKEATH
Howard Hughes Institute, Laboratory of Molecular Biology, University of Wisconsin-Madison, 1525 Linden Drive, Madison, WI 53706, USA.

BRUCE S. THALLEY
Howard Hughes Institute, Laboratory of Molecular Biology, University of Wisconsin-Madison, 1525 Linden Drive, Madison, WI 53706, USA.

JIM A. WILLIAMS
Howard Hughes Institute, Laboratory of Molecular Biology, University of Wisconsin-Madison, 1525 Linden Drive, Madison, WI 53706, USA.

RICHARD A. YOUNG
Whitehead Institute for Biomedical Research, Nine Cambridge Center, Cambridge, Massachusetts 02142; and Department of Biology, Massachusetts Institute of Technology, Cambridge, Massachusetts 02139, USA.

Abbreviations

AcMNPV	MNPV from *Autographa california*
ARS	autonomous replication sequence
BmNPV	NVP from *Bombyx mori*
BSA	bovine serum albumin
BV	budded virus
cDNA	complementary DNA
CEN	yeast centromeric sequence
c.f.u.	colony-forming units
CIP	calf intestinal phosphatase
CM	carboxymethyl
DEAE	diethylaminoethyl
DEPC	diethylpyrocarbonate
DMEM	Dulbecco's modified Eagle's medium
DMSO	dimethylsulfoxide
DTT	dithiothreitol
ECV	extracellular virus particles
EDTA	ethylenediamine tetraacetic acid
EGF	epidermal growth factor
ELISA	enzyme-linked immunosorbent assay
Endo H	endoglycosidase H
FCS	fetal calf serum
FPLC	fast protein liquid chromatography
GST	glutathione S-transferase
HA	hypoxanthine/azaserine
HAT	hypoxanthine/aminopterin/thymidine
HSA	human serum albumin
IDA	iminodiacetic acid
IgG	immunoglobulin G
IPTG	isopropyl-β-D-thiogalactopyranoside
MBP	maltose binding protein
MCAC	metal chelate affinity chromatography
MNPV	multiple virion NPV
m.o.i.	multiplicity of infection
NPV	nuclear polyhedrosis virus
NTA	nitrilotriacetic acid
ORF	open reading frame
OV	occluded virus
PBS	phosphate-buffered saline
PCR	polymerase chain reaction

PEG	polyethylene glycol
p.f.u.	plaque-forming units
PGK	phosphoglycerate kinase
p.i.	post-infection
PMSF	phenylmethylsulfonyl fluoride
QAE	quaternary aminoethyl
RIA	radioimmunoassay
SDS	sodium dodecyl sulfate
SDS–PAGE	SDS–polyacrylamide gel electrophoresis
Sf9	cell line used in baculovirus studies
SH2	src homology 2
SH3	src homology 3
SNPV	single virion NPV
SP	sulfopropyl
SPA	Staphylococcal protein A
TCA	trichloroacetic acid
TFA	trifluoroacetic acid
UAS	upstream activating sequence
UTR	untranslated region
X-Gal	5-bromo-4-chloro-3-indolyl-β-D-galactopyranoside

1

Screening λ expression libraries with antibody and protein probes

BENJAMIN MARGOLIS and RICHARD A. YOUNG

1. Introduction

The λgt11 vector–host system and related gene expression systems permit the isolation of specific recombinant DNA sequences by virtue of the antigenicity or activity of the expressed gene product. The expressed protein, or domain of the protein, can be detected by using antibodies or through binding to a specific substrate such as a DNA fragment or a polypeptide. The design of the λgt11, λZAP, and λEXlox systems have been described in detail elsewhere (1–4). We describe here features of the recombinant DNA libraries and the antibody and the protein probes that are important for successful screening. We also describe our favourite screening protocols using antibody and protein probes.

2. The DNA library and expression of foreign DNA

The quality of the recombinant DNA library and the quality of the probe are the major determinants of the frequency with which positive signals are obtained. A 'good' library is large enough to contain multiple copies of the genomic DNA or cDNA of interest. Detailed protocols for the construction of cDNA and genomic DNA libraries in λgt11 and related vectors are published elsewhere (5–7), and λgt11 libraries are commercially available (Clontech).

There are three major problems associated with obtaining expression of foreign DNA as a stable polypeptide. The first problem is that most foreign DNA does not contain the transcription control signals required for expression in *E. coli*. The foreign gene must be placed under the control of an *E. coli* promoter that is efficiently recognized by *E. coli* RNA polymerase. The second problem is that unusual polypeptides are often rapidly degraded in *E. coli* (8–11). In some cases, the instability of foreign proteins can be reduced by fusing the antigen to a stable host protein and by using host mutants deficient in proteolysis (1, 12–14). The third major problem with foreign protein synthesis in *E. coli* is that the presence of these unusual proteins is often harmful or even lethal to the cell. Demanding high levels of gene expression can compound this problem, since constitutive high level expression of even normal components of the cell can often be lethal (15). One solution to this problem has been to ensure that expression of the foreign protein is transient; the expression of the DNA encoding the foreign protein is repressed during early log phase growth of the host cell culture, and the expression of the foreign protein is induced near the end of this period when the transcriptional and translational apparatus are still fully active.

To maximize the levels to which foreign proteins can accumulate in *E. coli*, these concepts were incorporated into the λgt11 expression vector–host system and several similar systems (1). Foreign DNA is inserted into these λ phage expression vectors under the control of regulated promoters. The recombinant λ phage are propagated lytically and expression of the foreign DNA is induced just before initiating the screening process.

3. Screening libraries with antibody probes

3.1 Antibody probes

The quality of the antibody probe will have an important influence on the success of a screen. It is often useful to investigate the behaviour of the antibody on a Western blot with the antigen of interest. Even highly specific antibodies sometimes bind to irrelevant proteins, albeit with much lower avidity. From a Western blot of a crude extract and purified protein, the investigator can learn the limits of antigen detection with the antibody, how many different proteins are bound by the antibody, the signal intensity obtained with the protein of interest relative to other cross-reacting proteins, and the experimental condition that produces the maximum signal-to-noise ratio for the antibody. Note that this information can be misleading, as it depends upon the relative amounts of the different proteins present in the crude extract. However, a Western blot will provide clues to the specificity of the antibody and the limitations that are inherent in the use of that antibody for screening an expression library.

Several factors influence the successful use of antibody probes of recombinant DNA expression libraries. The amount of antigen that associated with the nitrocellulose filter during a plaque lift varies with each fusion protein, and is related to its level of expression and stability in the host cells. For λgt11, successful screens with antibody probes have been accomplished with plaques containing as little as 30–60 picograms of antigen, although many plaques contain as much as 10–30 nanograms of recombinant protein. The ability to detect limited amounts of antigen will depend upon the titre and the binding characteristics of the antibody. The best signal-to-noise ratios are produced by high titre, high affinity antibodies used at low dilution (approximately 1:1000). Antibodies that produce good signals on Western blots generally produce good signals in the λgt11 screening procedure when used at similar dilutions.

Both polyclonal and monoclonal antibodies have been used successfully to isolate λgt11 clones. Since polyclonal antibodies often recognize multiple determinants, they may identify a larger fraction of the set of all clones that express a protein of interest. Both monoclonal and polyclonal antibodies can bind multiple proteins by virtue of cross-reactive epitopes, and isolation of the correct DNA must be confirmed by additional criteria.

3.2 Screening procedure

This is described fully in *Protocol 1*. Two methods for the detection of the bound antibody are described in this procedure, incorporating either horse-radish peroxidase-coupled or alkaline phosphatase-coupled second antibody. These techniques produce signals directly on the nitrocellulose filter, thereby exactly reproducing the pattern of plaques on the plate (a faint background is produced by each plaque). This allows the precise location of the single plaque producing the signal, reducing the subsequent work involved in plaque purification. Moreover, the frequency of false positives is very low, and these differ in appearance from the doughnut shaped, plaque sized genuine positive signals. Signal-to-noise ratios and sensitivities obtained by using the two methods (described in *Protocol 1*) do not differ substantially. The first of the two techniques described (*Protocol 1C*) is less time-consuming.

Protocol 1. Screening with antibody probes

Equipment and reagents

- λgt11 library
- *E. coli* Y1090
- LB broth: 10 g/litre Bacto tryptone, 5 g/litre Bacto yeast extract, 10 g/litre NaCl, pH 7.5
- LB containing 0.2% maltose
- LB plates (90 mm or 150 mm in diameter): LB containing 15 g/litre Bacto agar
- LB plates containing 50 μg/ml ampicillin
- LB soft agar: LB containing 8 g/litre Bacto agar
- λ diluent: 10 mM Tris–HCl pH 7.5, 10 mM MgCl₂
- 10 mM IPTG

- TBS: 50 mM Tris–HCl pH 8.0, 150 mM NaCl
- TBST: 50 mM Tris–HCl pH 8.0, 150 mM NaCl, 0.05% Tween-20
- TBST plus 0.5% BSA
- TBST with 0.1% BSA and primary antibody
- Detection kit [e.g. alkaline phosphatase-conjugated second antibody (Promega Biotec), or biotinylated antibody plus avidin-conjugated horse-radish peroxidase (Vector Laboratories)]
- Nitrocellulose filters (Schleicher and Schuell BA85, use fresh)

A. Plating the library

1. Streak *E. coli* Y1090 for single colonies on LB plates (pH 7.5) containing 50 μg/ml ampicillin, and incubate at 37 °C.

2. Beginning with a single colony, grow Y1090 to saturation in LB plus 0.2% maltose at 37 °C with good aeration. Add MgSO₄ to 10 mM.

Protocol 1. *Continued*

3. For each 90 mm LB plate, mix 0.2 ml of the Y1090 culture with 0.1 ml of λ diluent containing up to 3×10^4 p.f.u. [a] of the λgt11 library. For 150 mm LB plates, use 0.6 ml of the Y1090 culture with up to 10^5 p.f.u. [b] in λ diluent. Adsorb the phage to the cells by incubation at 37 °C for 15 min.

4. Add 2.5 ml (for a 90 mm plate) or 7.5 ml (for a 150 mm plate) LB soft agar to the culture and plate. Incubate the plates at 42 °C for 3–4 h. Plaques should be just visible at this point.

B. *Screening the library*

1. Move the plates to a 37 °C incubator. Quickly overlay each plate with a damp nitrocellulose filter disk which has been wetted in 10 mM IPTG. Schleicher and Schuell BA85 filters are particularly good for this protocol. Fresh filters are ideal, as very old filters may lose their binding properties. If necessary, blot the nitrocellulose to ensure that no pools of liquid remain before overlaying on the plate. Incubate for 3–4 h longer at 37 °C.

2. Move the plates to room temperature, mark the position of the filter on the plate with a needle, and remove the filters carefully. If the top agar tends to stick to the filter rather than the bottom agar, chill the plates at 4 °C for 10–20 min.

   Do not allow the filters to dry out during any of the subsequent steps. Perform all of the following washing and incubation steps at room temperature with gentle shaking.

3. Rinse the filters briefly in TBST.

4. Incubate the filters, antigen side up, in TBST plus 0.5% BSA for 5 min. Use 5 ml per 82 mm filter, and 10 ml per 132 mm filter.

5. Incubate the filters in TBST with 0.1% BSA and antibody with gentle shaking for 1 h at room temperature. Use 5 ml per 82 mm filter, and 10 ml per 132 mm filter.

6. Wash the filters in TBST three times, 5 min each time.

C. *Detection of bound antibody using alkaline phosphatase* (Promega Biotec)

1. Transfer the washed, antibody-bound filters to TBST containing affinity purified, alkaline phosphatase-conjugated goat IgG (5 µl antibody/5 ml of TBST). Incubate for 30 min with gentle agitation.

2. Wash the filters in TBS (no Tween-20) three times, 5 min each time.

3. Incubate the filters (antigen side up) with the substrate solution according to the manufacturer's instructions.

4. After the colour develops, wash with two changes of distilled water, and allow to dry.

D. *Detection of bound antibody using horse-radish peroxidase* (Vector Laboratories)

1. Transfer the filters to TBST containing biotinylated second antibody, used according to the manufacturer's instructions.

2. Wash the filters in TBST three times, 5 min each time.

3. Transfer the filters to TBST containing Vectastain ABC reagent. Incubate them (antigen side up) for 30 min with gentle agitation.

4. Wash in TBS (no Tween-20) three times, 5 min each time.

5. Incubate the filters in the peroxidase substrate solution. After the colour develops, wash with two changes of distilled water, and allow to dry.

[a] The authors use 10^4 p.f.u. at this step.
[b] The authors use 3×10^4 p.f.u. at this step.

3.3 Comments on the screening procedure

(a) Using a fresh overnight of *E. coli* Y1090 will minimize the number of non-viable cells and the lag time.

(b) The plaque size can be adjusted by reducing or increasing the amount of cells plated.

(c) The plates used should be fresh because phage diffusion will be greatest in moist plates, producing large plaques, which is desirable. However, the excessive moisture in fresh plates will pool on the agar surface once in the incubator, and plaques will smear. The best solution is to plate cells on fresh (0–2 day-old) plates, place them in the 42 °C incubator, and then check for pools of liquid after 15 minutes. The tops of plates with pooled liquid can be tilted ajar for a few minutes to permit evaporation. *Do not overdry*. Pools can remain on the surface of plates for up to 40 minutes without causing smearing of the plaques.

(d) Plaques should just be visible before overlaying the nitrocellulose filter. If λgt11 plaques are not clearly visible within four hours after plating, there is a problem with the cells, phage, or plates.

(e) If top agar sticks to the nitrocellulose filters, even when the filters are removed carefully and slowly, and chilling the plates at 4 °C does not eliminate the problem, do not despair. Although it will be more difficult to find a specific plaque on the plate if the top agar fails to remain intact, large numbers of phage remain on the surface of the bottom agar after the top agar is removed, allowing the isolation of phage in the area of a plaque.

(f) During the washing and blocking step, almost any source of protein can be substituted for BSA, if desired, as long as it is not bound by antibody.

(g) If necessary, the filter can be incubated with antibody overnight without significant changes in signal. However, phage continue to diffuse in the agar during this time which may affect the level of clonal purity, particularly in later screens.

(h) The antibody binding conditions can be changed if preliminary experiments with Western blots indicate that alternate conditions improve the signal-to-noise ratio.

(i) Many investigators choose to rock the filters back and forth during antibody binding and washing steps. This is not necessary but it helps somewhat. Agitating the filters in a circular motion often produces undesirable effects, as reagents tend to concentrate around the outer areas of the filter.

(j) With protocols using non-radioactive antibody probes, signals should be obtained within 15 minutes and are sometimes seen within only one to two minutes after the addition of the substrate solution. Protocols which use $[^{125}I]$-protein A (1, 2), while sensitive, are prone to producing false positive signals and so are often abandoned by most investigators.

(k) Variations on the screening method can be found in numerous sources, for example, Snyder *et al.* (16).

4. Screening libraries with protein probes

Bacterial expression libraries can be used to identify not only antibody–antigen interaction but also other forms of protein interactions. In this approach, labelled proteins are substituted for antibodies and used to probe λgt11, λZAP, or λEXlox libraries. Many laboratories have exploited this system to examine intracellular protein interactions in mammalian cells (e.g. 17–22). This technique is quicker and simpler than the conventional approach of purifying binding proteins by affinity chromatography, obtaining protein sequence, and then cloning using degenerate oligonucleotides.

This section will discuss the use of bacterial expression libraries to clone genes using protein–protein interactions. We first discuss important considerations for protein probes, then we describe a procedure for screening with protein probes, including:

- how to prepare protein probes
- how to handle λEXlox libraries (since they are handled differently than λgt11 or λZAP)
- how to screen expression libraries with protein probes.

4.1 Protein probes

Prior to using this approach, it is helpful to have some preliminary understanding of the nature of the protein interactions. The method will not be successful if the protein expressed in the phage requires modification for binding activity. For this reason, this approach seems to work best for intracellular protein interactions rather than extracellular interactions where complex folding and carbohydrate or lipid modification may be essential for binding. However, this method may also fail to detect certain intracellular protein interactions that are of a low affinity or that are dependent on post-translational modifications.

One of us (B. Margolis) has utilized bacterial expression libraries to clone src homology 2 (SH2) domain proteins that interact with the tyrosine phosphorylated carboxy terminal tail of the EGF-receptor (19, 20). A system was developed which allowed *in vitro* labelling of the EGF-receptor tyrosine phosphorylation sites with $[\gamma-^{32}P]ATP$. This labelled receptor was then used to screen bacterial expression libraries and resulted in the cloning of several new SH2 domain proteins. In the case of the SH2 domain–EGF-receptor interaction, the receptor must be tyrosine phosphorylated to bind the SH2 domain protein. The method is successful only if one uses the tyrosine phosphorylated probe to clone the SH2 domain protein, because *E. coli* cells do not contain tyrosine kinases.

There are examples of using bacterial expression cloning to delineate protein interactions even when the molecular basis of the interaction is not understood. This approach was taken by Baltimore and co-workers who used src homology 3 (SH3) domains to screen λgt11 libraries despite the fact that little was understood about the role of SH3 domains in these types of interactions (23). Nonetheless, this group was able to clone SH3 domain binding proteins and to characterize the binding partners for SH3 domains as proline-rich sequences. Thus, an understanding of the protein interaction is helpful but not always essential.

4.2 Preparation of protein probes

Probes can be labelled in a variety of ways depending on whether the probe is a cloned gene or a purified protein. If only purified protein is available, then one can use standard iodination or biotinylation techniques. Iodination tends to be labour intensive, can modify protein structure, and requires that a fumehood suitable for radioactive work is available. Biotinylation is very convenient and has been used by some researchers (23), but in our experience is not as sensitive as ^{32}P-based detection. The use of a glutathione S-transferase (GST) bacterial fusion protein (see Chapters 2 and 3) seems the best choice if the gene for the probe has been cloned. These fusion proteins can incorporate a protein kinase A phosphorylation site which allows labelling of the probe with ^{32}P (21, 22).

To label GST fusion proteins with protein kinase A, we use the procedures described in *Protocol 2*, which is adapted from Ron and Dressler (24). Proteins labelled in this fashion can be used to probe λgt11, λZAP or λEXlox libraries.

Protocol 2. Preparation of ^{32}P-labelled GST fusion proteins as probes

Equipment and reagents

- cDNA encoding the protein probe
- pGEX-2TK vector (Pharmacia)
- Glutathione agarose (Sigma G4510)
- Elution buffer: 50 mM Tris–HCl pH 8.0, 1 mM EDTA, 0.5 mg/ml BSA, 0.005% NP-40, 1 mM DTT, 1 mM PMSF, 10 mM glutathione[a]
- Sephadex G-50 (coarse)
- DK buffer: 50 mM potassium phosphate buffer pH 7.15, 10 mM MgCl₂, 5 mM sodium fluoride, 4.5 mM DTT
- Reaction buffer. This is composed of 0.2 U/ μl of protein kinase A catalytic subunit (Sigma P2645) and 750 μCi [γ-^{32}P]ATP (Dupont/ NEN, NEG-35C, 6000 Ci/mmol, 150 mCi/ml,

unpurified) added to DK buffer. The protein kinase A powder as supplied by the manufacturer should be aliquoted into small fractions of ~ 0.5–1.0 mg (10–20 U depending on the specific activity of the material supplied) and frozen dry at −20°C. 10 min prior to use, resuspend the powder in 10 μl deionized water containing 6 mg/ml DTT. A typical reaction volume of 50 μl consists of 6 μl protein kinase A (10 U), 5 μl of [γ-^{32}P]ATP, and 39 μl of DK buffer.
- PBS containing 1 mM DTT
- Reagents and apparatus for SDS–PAGE

Method

1. Subclone the cDNA encoding the protein probe into a pGEX vector (see Chapter 2) to encode an in-frame GST fusion protein containing a protein kinase A site. Such vectors are available from Pharmacia (pGEX-2TK) or from the authors of ref. 24.

2. Use standard techniques to generate the GST fusion proteins in *E. coli* and bind the fusion protein to glutathione agarose (25, 26, and Chapter 3). The protein generated must be at least partially soluble to allow purification on glutathione agarose.

3. Use SDS–polyacrylamide gel electrophoresis (SDS–PAGE) to quantitate the amount of protein bound to the beads and the amount that can be eluted. A standard technique for elution is described in step 9 and should be attempted before radiolabelling.[a] To generate a good probe, one should be able to elute at least 50% of the bound fusion protein. To quantitate the amount of protein bound to the beads boil the beads in 1 × SDS–PAGE sample buffer. To quantitate the amount eluted, add the appropriate volume of 3 × SDS–PAGE sample buffer to the eluted sample and boil. Run both the beads before elution and the protein eluted from the beads on an SDS gel. Run different quantities of BSA ranging from 5–40 μg as standards. Stain the gel with Coomassie blue to quantitate the amount of protein originally bound to the beads and the amount eluted using BSA as a standard.

4. Place the appropriate volume of glutathione agarose bearing 10 μg of bound protein in a microcentrifuge tube. Add coarse pre-swelled Sephadex G-50 or other agarose beads to give a 50% slurry volume of at least 40 μl.

5. Wash the beads twice with 1 ml of DK buffer.

6. Incubate the beads with 50 μl of freshly prepared reaction buffer.

7. Shake the phosphorylation reaction for 30 min at 30 °C.

8. Remove the unincorporated [γ-^{32}P]ATP by washing the beads three times with PBS containing 1 mM DTT.

9. Elute the phosphorylated fusion protein from the beads with 400 μl of freshly prepared elution buffer by shaking for 30 min at room temperature.

10. Spin the beads for 2 min in a microcentrifuge and remove the supernatant which should contain the phosphorylated probe (this can be checked with a Geiger counter).

11. Add another 400 μl of elution buffer, briefly mix, and spin down. Pool the two supernatants together.

12. Check the labelling efficiency by counting 2 μl; it should be approximately 1×10^7 d.p.m. μg.

[a] Certain proteins elute better at an alkaline or acidic pH without loss of binding activity. We have used 0.1 M sodium borate pH 9.0 or 0.1 M sodium citrate pH 6.0 to elute proteins in some cases.

4.3 Screening procedure

This section focuses on the use of λEXlox, a variant of λgt11, and differences in the procedure for plating this phage and λgt11. This phage was described by Pallazallo *et al.* (4) and is based on the pET expression systems developed by Studier and co-workers (27). In this system, cDNA clones are fused to a fragment of the T7 capsid protein gene 10 under the control of the T7 promoter. These phages are used to infect *E. coli* harbouring the T7 polymerase under *lacUV5* control. Induction with IPTG generates the T7 polymerase which then initiates transcription of the fusion protein encoded by the phage library. T7 RNA polymerase is more powerful than the normal *E. coli* RNA polymerase and therefore expression of plaque protein tends to be higher. λEXlox also provides the advantage of automatic subcloning from phage to plasmid by passing the purified phage through a bacterial strain containing the P1 Cre recombinase. The method for construction and plating are similar to λgt11 except that the bacterial strains which harbour the T7 polymerase are more difficult to use in library plating.

All components for the preparation and use of λEXlox libraries are available from Novagen which also sells several high quality libraries.

Protocol 3. Screening with protein probes

Equipment and reagents

- *E. coli* BL21 (DE3) pLysE
- λEXlox library
- 2×YT: 16 g tryptone, 10 g yeast extract, 5 g NaCl per litre
- 2×YT containing 0.2% maltose, 10 mM MgSO$_4$, and 25 μg/ml chloramphenicol
- 2×YT plates (150 mm diameter plates): 2×YT containing 15 g agar and 25 μg/ml chloramphenicol
- LB soft agarose: LB containing 0.7% agarose (see *Protocol 1* for composition of LB)
- SM: 50 mM Tris–HCl pH 7.8, 8 mM MgSO$_4$, 100 mM NaCl, 0.1% gelatin
- Nitrocellulose filters (*Protocol 1*)
- 1 mM IPTG
- TBS containing 0.1% Triton X-100 (see *Protocol 1* for composition of TBS)

- Block buffer: 20 mM Hepes pH 7.5, 5 mM MgCl$_2$, 1 mM KCl, 5 mM DTT, 5% non-fat dry milk, 0.02% sodium azide
- ^{32}P-labelled protein probe (from *Protocol 2*). Before adding the probe to the block buffer, filter the probe using a 1 ml syringe and a syringe filter (Millipore SJHVOO4NS). Perform this step slowly so as not to dissociate the syringe and filter due to pressure build-up. This step is important to eliminate background spots (perhaps aggregates of the probe) which can lead to false positives.
- Crystallizing dishes (150 mm diameter and 170 mm diameter; Pyrex 3140)
- X-ray film and developer

A. *Plating the library*

1. Streak out a plate of BL21(DE3)pLysE on 2×YT plates containing 25 μg/ml chloramphenicol.

2. For each plating of the library, pick a new colony of this bacterium and grow at 37 °C in 2×YT with 0.2% maltose, 10 mM MgSO$_4$, and 25 μg/ml chloramphenicol. These bacteria grow slowly, so allow at least 16 h for the bacteria to reach saturation. Problems with BL21(DE3)pLysE are the major cause for failure of library plating.

3. In a 15 ml tube, mix 100 μl of saturated bacteria, 100 μl 10 mM MgSO$_4$/10 mM CaCl$_2$, and 100 μl of 4 × 10^4 phage diluted in SM.

4. Shake for 30 min at 37 °C.

5. Add 10 ml of LB soft agarose pre-warmed to 50 °C to the bacteria. Quickly mix and pour onto a 150 mm 2×YT plate containing 25 μg/ml chloramphenicol which is dry and pre-warmed (as described in Section 3.3) to 37 °C.

6. Allow plates to grow 8–12 h at 37 °C until the plaques are touching but not yet confluent. The longer plating time compared to λgt11 (*Protocol 1*) is due to the slow growth of the bacteria.

B. *Screening the library*

1. Overlay each plate with a 137 mm nitrocellulose filter impregnated with IPTG. To prepare these filters, wet the nitrocellulose circles with 1 mM IPTG in water and then dry on aluminium foil for 1 h. Label the filters with a black ballpoint pen before placing on plates.

2. With the filters on the plate, mark the orientation by puncturing the edge of the filters in four to five different positions with a 21 gauge needle. By placing the plates on a light box, the positions of the needle holes can then be marked on the bottom of the plates. Leave the filter on the plate overnight at 37 °C. It is not possible to do duplicate lifts.

3. In the morning, remove the filters and place in TBS containing 0.1% Triton X-100. The plaques should still be clear although unlike λgt11 they may shrink slightly in size after the IPTG treatment. Store the plates at 4 °C.

4. Wash the filters three times with TBS containing 0.1% Triton using a 170 mm crystallizing dish (Pyrex 3140). Some workers denature and renature the proteins at this point (28) but we have not found this to be necessary. Unless noted, it is important that the filters be transferred one at a time from one solution to the next during all washes and incubations. This prevents the filters from sticking together which can lead to background problems.

5. Use a 150 mm crystallizing dish (Pyrex 3140) for blocking and incubation. Incubate the filters for at least 3 h in block buffer at 4 °C. Filters can be stored in this block buffer for at least a week at 4 °C. Use a 150 mm crystallizing dish for blocking and incubation and a 170 mm crystallizing dish for washing.

6. Discard the block buffer and add the radiolabelled probe which has been filtered (see *Equipment and reagents*) to new block buffer (final concentration 2×10^6 d.p.m./ml). Incubate the filters with the probe at 4 °C overnight. We probe ten filters at a time by incubating in 60 ml of buffer.[a]

7. Remove the probe and save for a second screening.[a]

8. Rinse the filters once with TBS containing 0.1% Triton without changing containers to remove the bulk of the unbound radioactivity.

9. Wash the filters in TBS containing 0.1% Triton four times for 15 min each time at room temperature.

10. Dry the filters and expose to X-ray film for 2–48 h at −70 °C with intensifying screens. The length of exposure that is required will depend on the probe radioactivity and the background.

11. Orient the film and the filters and mark the position of the needle holes on the film. Then align the plates and the film using the needle holes and mark the area of the plates that contain the positive signal.

C. *Isolation of positive plaques*

1. Pick the positive plaques using the non-tapered cotton-plugged end of a sterile 5 ml pipette (e.g. Falcon 7543).

2. Place the plug containing the plaques into a microcentrifuge tube

Protocol 3. *Continued*

> containing 1 ml of SM. Shake for 2 h at room temperature to release the phage from the plug.

3. Dilute the phages 1:10 000 with fresh SM. Using 100 μl of the diluted λEXlox phage, re-plate the primary positives on 90 mm or 150 mm dishes, and repeat parts A and B.

4. Select the resultant positives and repeat the steps to purify the phage to homogeneity. With each successive purification, pick the positives with smaller diameter pipettes to ensure purification of a single clone. Once the phage is purified, DNA can be isolated for sequencing.

[a] You can save the probe and reuse it a second time. However it is best to use fresh probe with each screening.

4.4 Comments on the screening procedure

(a) A positive control is very helpful in establishing these techniques. In the case of the EGF-receptor/SH2 domain binding, it was well known before the screening that the EGF-receptor bound to the SH2 domain containing protein, phospholipase C-γ (PLC-γ). Thus, as a control for the EGF-receptor probe, PLC-γ was spotted on nitrocellulose and then probed as described above. Two nanograms of spotted protein could be easily detected using this approach. One should aim for a probe that can detect at least ten nanograms of control protein. An even better control is to subclone the cDNA for a known binding protein or domain into λgt11 or λEXlox. The SH2 domain of PLC-γ was subcloned into λEXlox and served as a control for both the probe and library plating (20). Obviously one must be very careful not to contaminate the library with the positive control.

(b) In a typical library screen of ten 150 mm plates, we typically obtain 10–20 positives. Of these, only 10–20% will be true positives on a secondary screen. This ratio may vary depending on the particular probe used but it is always safer to pick all positives rather than guess at which ones appear to be true positives. However, if one has a high background (e.g. 10 positives per plate) it becomes impossible to pick all the plaques to separate true from false positives. In this case, it is better to find the cause of the high background before proceeding with further screens.

(c) It is important to note that the titres of phage obtained from λEXlox infection of BL21(DE3)pLysE are typically 100-fold lower than those obtained with λgt11 infection of Y1090.

(d) After the phage is purified, λEXlox can be converted to a plasmid by passing the phage through the bacterium BM25.8. This bacterium and protocols for its use are also available from Novagen. We take 100 μl of a 10^2–10^4 dilution of a purified phage (depending on titre) and mix this with 100 μl of BM25.8 bacteria to obtain bacteria containing the plasmid. The plasmid DNA obtained from mini-preps of BM25.8 is not suitable for analysis and must be used to re-transform a bacterial strain such as DH5α (other strains such as HB101 work just as well). Plasmids obtained from these strains can then be used for restriction analysis and double-stranded sequencing.

(e) The final plasmids obtained can also be used to re-transform competent BL21(DE3)pLysE bacteria. These bacteria can then be induced with IPTG to produce a fusion protein of T7 gene 10 linked to the newly cloned gene (27). Expression of the fusion protein is high but the protein is usually insoluble. We prefer to cut the insert from the DH5α plasmid DNA (or obtain it by PCR) and ligate into a pGEX vector to generate an in-frame GST fusion protein. The protein obtained is usually partially soluble and can be used for cellular protein interaction studies after purification on glutathione agarose. Such GST fusion proteins also make good immunogens.

(f) λEXlox can also be used for nucleic acid screening by plating with the bacterial strain ER1647 instead of BL21(DE3)pLysE. This strain is plated on LB agar plates without antibiotics and the plates are grown for 6–8 hours rather than the 8–12 hours with BL21(DE3)pLysE. ER1647 is easier to work with and yields a higher phage titre from plaques than BL21(DE3)pLysE.

Acknowledgements

B. M. wishes to thank E. Skolnik, C. Roonprapunt, E. Lowenstein, M. Jaye, B. Morris, P. Olivier, and V. J. Yajnik for help in developing these protocols. We thank D. Ron for the pGEX plasmid with protein kinase A site.

References

1. Young, R. A. and Davis, R. W. (1983). *Proc. Natl. Acad. Sci. USA*, **80**, 1194.
2. Young, R. A. and Davis, R. W. (1983). *Science*, **222**, 778.
3. Short, J. M., Fernandez, J. M., Sorge, J. A., and Huse, W. D. (1988). *Nucleic Acids Res.*, **16**, 7583.
4. Palazzalo, M. J., Hamilton, B. A., Ding, D., Martin, C., Mead, D. A., Mierendorf, R. C., Raghavan, K. J., Meyerowitz, E. M., and Lipshitz, H. D. (1990). *Gene*, **88**, 25.
5. Young, R. A., Bloom, B., Grosskinsky, C., Ivanyi, J., Thomas, D., and Davis, R. W. (1985). *Proc. Natl. Acad. Sci. USA*, **82**, 2583.
6. Huynh, T. V., Young, R. A., and Davis, R. W. (1985). In *DNA cloning techniques: a practical approach* (ed. D. Glover), Vol. I, pp. 49–78. IRL Press, Oxford.

7. Sambrook, J., Fritsch, E. F., and Maniatis, T. (ed.) (1989). In *Molecular cloning, a laboratory manual* (2nd edn), pp. 8.3–8.82. Cold Spring Harbor Laboratory Press, Cold Spring Harbor, NY.
8. Goldberg, A. L. and St. John, A. C. (1976). *Annu. Rev. Biochem.*, **45**, 747.
9. Charnay, P., Gervais, M., Louise, A., Galibert, F., and Tiollais, P. (1980). *Nature*, **286**, 893.
10. Edman, J. C., Hallewell, R. A., Valenzuela, P., Goodman, H. M., and Rutter, W. J. (1981). *Nature*, **291**, 504.
11. Kupper, H., Keller, W., Jurtz, C., Forss, S., Schaller, H., Franze, R., Strommaier, K., Marquardt, O., Zaslavsky, V. G., and Hofschneider, P. H. (1981). *Nature*, **289**, 555.
12. Itakura, K., Hirose, T., Crea, R., Riggs, A. D., Heyneker, H. L., Bolivar, F., and Boyer, H. W. (1977). *Science*, **198**, 1056.
13. Goeddel, D. V., Kleid, D. G., Bolivar, F., Heyneker, H. L., Yansura, D. G., Crea, R., Hirose, T., Kraszewski, A., Itakura, K., and Riggs, A. D. (1979). *Proc. Natl. Acad. Sci. USA*, **76**, 106.
14. Davis, A. R., Nayak, D. P., Ueda, M., Hiti, A. L., Dowbenko, D., and Kleid, D. G. (1981). *Proc. Natl. Acad. Sci. USA*, **78**, 5376.
15. Shatzman, A., Ho, Y.-S., and Rosenberg, M. (1983). In *Experimental manipulation of gene expression*, pp. 1–16. Academic Press, New York.
16. Snyder, M., Sweetser, D., Young, R. A., and Davis, R. W. (1987). In *Methods in enzymology* (eds R. Wu, L. Grossman), Vol. 154, pp. 107–28. Academic Press, San Diego.
17. Sri Widada, J., Asselin, J., Colote, S., Marti, J., Ferraz, C., Trave, G., Haiech, J., and Liautard, J. P. (1989). *J. Mol. Biol.*, **205**, 455.
18. Bregman, D. B., Bhattacharyya, N., and Rubin, C. S. (1989). *J. Biol. Chem.*, **264**, 4648.
19. Skolnik, E. Y., Margolis, B., Mohammadi, M., Lowenstein, E., Fischer, R., Drepps, A., Ullrich, A. and Schlessinger, J. (1991). *Cell*, **65**, 83.
20. Margolis, B., Silvennoinen, O., Comoglio, F., Roonprapunt, C., Skolnik, E., Ullrich, A., and Schlessinger, J. (1992). *Proc. Natl. Acad. Sci. USA*, **89**, 8894.
21. Ron, D. and Habener, J. F. (1992). *Genes Dev.*, **6**, 439.
22. Kaelin, W. G. Jr., Krek, W., Sellers, W. R., DeCaprio, J. A., Ajchenbaum, F., Fuchs, C. S., Chittenden, T., Li, Y., Farnham, P. J., Blanar, M. A., Livingston, D. M., and Flemington, E. K. (1992). *Cell*, **70**, 351.
23. Cicchetti, P., Mayer, B. J., Thiel, G., and Baltimore, D. (1992). *Science*, **257**, 803.
24. Ron, D. and Dressler, H. (1992). *BioTechniques*, **13**, 866.
25. Smith, D. B. and Corcoran, J. M. (1993). In *Current protocols in molecular biology* (eds F. M. Ausubel, R. Brent, R. E. Kingston, D. D. Moore, J. G. Seidman, J. A. Smith, and K. Struhl), pp. 16.7.1–16.7.8. John Wiley and Sons, New York.
26. Frangioni, J. V. and Neel, B. G. (1993). *Anal. Biochem.*, **210**, 179.
27. Studier, F. W., Rosenberg, A. H., Dunn, J. J., and Dubendorff, J. W. (1992). In *Methods in enzymology.* (ed. D. V. Goeddel), Vol. 185, pp. 60–94. Academic Press, San Diego.
28. Macgregor, P. F., Abate, C., and Curran, T. (1990). *Oncogene*, **5**, 451.

2

Expression of foreign proteins in *E. coli* using plasmid vectors and purification of specific polyclonal antibodies

JIM A. WILLIAMS, JAMES A. LANGELAND, BRUCE S. THALLEY, JAMES B. SKEATH, and SEAN B. CARROLL

1. Introduction

For several decades antibodies have been used to identify, localize, quantitate, and analyse proteins. Before the advent of gene cloning, only natural sources were available for the isolation of proteins to which antibodies were raised. Low protein yields and impurities often compromised the quality and quantity of antibodies that could be produced. In addition, the vast majority of proteins that play critical roles in regulating cellular processes were completely unknown. As powerful genetic approaches have been devised to isolate the genes involved in a vast array of biological processes in diverse organisms, most proteins are now first identified at the gene level. The analysis of the structure, expression, regulation, and function of proteins identified in this manner usually requires the production of specific antibodies that can be used to detect the protein in different cells, tissues, organisms, and other biological samples.

The production of proteins of limited natural availability has been made possible in both prokaryotic and eukaryotic host systems by the development of a wide variety of vectors for the efficient transcription of cloned DNA and the expression of heterologous proteins. The maximum yield and biological activity of individual recombinant proteins varies greatly and depends upon a large number of factors (e.g. protein stability, solubility, size, post-translational modifications, etc.) that often differ between expression systems (1–3). In general, however, the immunogenicity and antigenicity of recombinant polypeptides are less dependent upon the host and expression system, and for this reason the wide variety of cleverly designed expression vectors available for use in *E. coli* has made these systems the most popular choices for producing recombinant proteins for immunological purposes.

There are two general strategies for expressing recombinant proteins to be used as antigens. The first involves the in-frame fusion of all or part of the DNA encoding the protein of interest (usually cDNA) to bacterial coding sequences such that a fusion protein is produced. In addition to providing a common molecular handle that can be used to purify the recombinant protein, fusion proteins may be more stable and in some cases more soluble in *E. coli* than the foreign protein alone. The most commonly used fusion vectors utilize β-galactosidase (4, 5), *trp E* (6), protein A (7), glutathione S-transferase (8), or bacteriophage T7 (9) coding sequences to facilitate high levels of protein expression. The second strategy involves expressing the foreign gene directly, usually in vectors providing some type of very high level inducible prokaryotic promoter and a leader sequence for efficient translation initiation such as those found in the pET family of vectors (9).

This chapter describes the use of some proven vector systems for the expression of proteins in *E. coli*, and techniques for the isolation of recombinant proteins and the production and purification of antibodies that recognize the native cognate of the recombinant protein. An emphasis is placed on generally useful techniques for protein manipulation and antibody production because, unlike protocols for recovering biological activity which are often empirical and protein-specific (see Chapter 3), the requirements for protein immunogenicity and antigenicity are less stringent and these protocols can be applied to virtually any protein produced in any expression system. For example, there is a common issue encountered in the high level expression of foreign proteins which is the formation of inclusion bodies containing aggregates of the recombinant protein of interest. Recovering such proteins in soluble form often involves denaturation/renaturation schemes (see Chapter 3) many of which are laborious, inefficient, and protein-specific (1, 2, 10). Here, we present some simple ways in which these initially insoluble aggregates can be manipulated to provide large quantities of antigen to be used for immunization and immunoaffinity chromatography. Overall, the chapter is intended to provide investigators with the rationale behind the different vector systems, animal hosts, and antibody purification schemes now available, and a detailed guide to the techniques that have been most successful with the largest number and widest variety of proteins.

2. Prokaryotic expression vectors

We provide here an overview of a variety of expression vectors that have been successfully used to express and purify recombinant proteins in *E. coli*. Due to design advantages and in-house experience, we recommend that initial expression strategies should utilize vectors from three of these systems, the pET, pGEX, and pMAL vectors. The other systems described here may be used as alternative vectors if difficulties arise expressing proteins in these recommended vectors.

2.1 pUR plasmid vectors

The pUR expression vector system has been described in detail in a previous chapter in this series (5). The basic feature of the system is that foreign DNA is cloned into the 3′ end of the E. coli lacZ gene resulting in a fusion protein that usually retains β-galactosidase activity. The pUR plasmids also contain the portion of pBR322 containing an origin of replication and the β-lactamase gene that confers ampicillin resistance and are maintained in host strains bearing the lacI^q allele that over-produce the lac repressor. β-Galactosidase fusion proteins are induced with isopropyl-β-D-thiogalactopyranoside (IPTG).

While exhibiting definite advantages over phage-based lacZ expression systems, a primary limitation of the pUR system is that the enormous size of the β-galactosidase protein (116 kd) is often far greater than that of the foreign segment of the fusion protein. Thus, one expresses and purifies a largely β-galactosidase antigen and the immune response to the protein of interest is not as great as it is when the foreign polypeptide constitutes the bulk of the antigen. For this reason, and perhaps due to additional differences in promoter strength, the pUR system has largely been supplanted by vectors that utilize either shorter fusion partners or none at all and are driven by powerful T7 or tac promoters.

2.2 pT7-7

This vector possesses the strong bacteriophage T7 promoter and is employed with a host strain in which expression of the T7 RNA polymerase, and hence expression of the foreign protein, is inducible by IPTG. However, this vector does not incorporate a fusion partner or other affinity tag to aid in purification of the foreign protein. A full description of vector pT7-7 and its application is given in Chapter 4.

2.3 pET expression vectors

The pET (plasmid for expression by T7 RNA polymerase) system vectors utilize the bacteriophage T7 RNA polymerase promoter to direct high level expression of cloned genes in E. coli (9) and are currently the prokaryotic system of choice for expressing foreign proteins. Studier et al. (9) present a comprehensive review of this system; only salient details will be discussed here.

The cloning cassette from the prototype expression vectors (pET-3a, 3b, and 3c) are shown in Figure 1b. The pET-3a–c series contains, within a short leader peptide, unique NdeI, NheI, and BamHI restriction enzyme sites that can be utilized to insert cloned sequences such that foreign open reading frames will be transcribed and translated under T7 polymerase control. The NdeI site fuses foreign inserts to the ATG start codon of the leader; cloning at this site removes any leader sequences from the translation product. Insertion

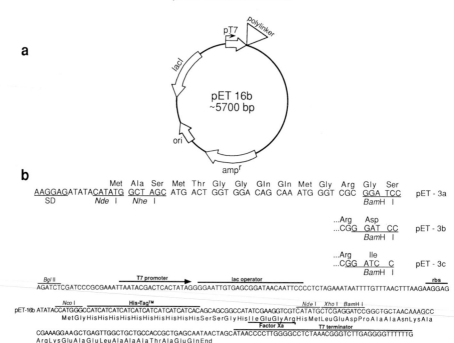

Figure 1. Structure of pET plasmids. (a) Location of cloning sites (polylinker) and structural features of the pET-16b plasmid; ori, origin of replication; amp^r, β-lactamase gene conferring ampicillin resistance; lacl, *lacl* gene; pT7, T7 promoter. (b) Sequence of the pET-3a cloning sites and unique restriction sites for cloning foreign DNA are shown, with modifications in the pET-3b and pET-3c vectors indicated. This region has been extensively modified in the pET-16b vector, shown at the *bottom* of the figure. The presence of the *lac* operator places the T7 promoter under *lac* control. Rbs denotes the ribosome binding sequence. His-Tag™ is a poly-histidine leader sequence. The factor Xa recognition sequence is underlined; factor Xa cleaves between the Arg and His of this sequence. The three unique cloning sites in the polylinker, *Nde*I, *Xho*I, and *Bam*HI, are indicated.

at the downstream *Bam*HI site results in translation of the insert fused to a short 12 amino acid leader peptide. The positions of the *Bam*HI sites are shifted with respect to the reading frame of the leader in pET-3a–c, such that any foreign coding region can be inserted at the *Bam*HI site of one of these vectors and be in the proper reading frame. An array of second generation pET expression vectors, based on these original constructs, have been constructed and are marketed by Novagen. These include expression vectors with extensive cloning cassettes (e.g. pET-17b), or the ability to express inserts as part of a larger fusion protein (pET-17xb). Some vectors contain modified T7–*lac* promoters and the *lac*I gene to reduce foreign protein expression from plasmids before induction (e.g. pET-16b, *Figure 1*); this modification is particularly useful when it is difficult to transform or maintain *E. coli* stocks of plasmids containing toxic inserts. Rather than the pET-3a–c leader peptide, the pET-16b vector also contains a poly-histidine encoding leader (*Figure 1b*) affinity tag. These histidine residues in the expressed protein bind to divalent ions (Ni^{2+}) allowing efficient affinity purification of the recombinant protein on nickel chelate columns. Bound protein is eluted with imidazole buffers and, if necessary, the recombinant protein can then be isolated from the poly-histidine tag by cleavage with protease factor Xa (see *Figure 1b*). Other pET vectors are available that contain C-terminal histidine repeats (e.g. pET-23a–c). Thus, when devising expression strategies for particular genes of interest, consultation of the maps and features of these other commercially available vectors may identify a more favourable vector.

All pET vectors share a number of common features, including the method-ology utilized for growth and expression in *E. coli*. All vectors contain the β-lactamase gene (conferring ampicillin resistance; Amp^r) as a selectable marker. A distinct advantage of this system is that the T7 RNA polymerase protein does not exist in standard laboratory stocks of *E. coli*. Thus all DNA manipulations may be performed in any RecA$^-$, K12 *E. coli* strain standardly used for cloning (e.g. HB101 or JM109), without concern that background expression of potentially toxic inserts will prevent recovery of the clone of interest. For protein expression, plasmids are transformed into a host strain containing a chromosomally integrated copy of the T7 RNA polymerase gene. This is present as a bacteriophage DE3 lysogen, and contains the T7 RNA polymerase gene under control of the *lacUV5* promoter (11). DE3 lysogens of BL21 (F$^-$ *ompT*r$_B^-$m$_B^-$) or HMS174 (F$^-$ *recA*r$_{K12}^-$m$_{K12}^+$Rif^r) are commonly used as expression host strains. In addition to lacking the *ompT* outer membrane protease, BL21 should also lack the *lon* protease (9). This may help stabilize foreign proteins that are sensitive to endogenous protease activity. Essentially any bacterial strain may be converted to a DE3 lysogen, utilizing a commercially available DE3 lysogenization kit (Novagen). In cases where the target protein is toxic, low level expression of the T7 RNA poly-merase gene in the BL21(DE3) strain may be sufficient to prevent stable transformation or propagation of the plasmid. This problem may be circum-vented in either of two ways. First, use of a pET vector containing the T7–*lac* promoter (e.g. pET-16b, *Figure 1b*) will place the foreign gene under *lac* control and thus reduce background target protein expression in the absence of IPTG. The alternative, more commonly utilized method, involves inhibi-tion of T7 polymerase protein produced in the DE3 lysogen via expression, from a second compatible plasmid, of small amounts of the T7 lysozyme gene. Two plasmids, conferring chloramphenicol resistance, carry this gene, and produce either low (pLysS) or high (pLysE) levels of lysozyme. Whenever possible, pLysS is utilized, since pLysE produces enough lysozyme to inhibit even induced levels of T7 polymerase, and significantly reduces protein induc-tion levels. If a protein is still too toxic to be expressed even utilizing pLysE, expression may be induced in the original cloning strain via infection with bacteriophage CE6, a lambda phage recombinant containing and expressing the T7 polymerase gene. Once the plasmid of interest is stably transformed into a DE3 lysogen, T7 expression from the *lacUV5* promoter, and subsequent foreign protein expression, is induced by adding IPTG to the culture.

2.4 pQE vectors

This series of vectors may be used as alternative histidine tagged vectors to allow affinity purification of fusion proteins on Ni^{2+} resin columns. Maps and vector details are available from the manufacturer (Hybaid Ltd).

2.5 Glutathione S-transferase (GST) expression vectors

While pET vectors are ideal for expressing native proteins, there are occasions where it is advantageous to express foreign proteins as gene fusions. For example, the use of carrier proteins that are easily purified via affinity chromatography greatly simplifies purification schemes required to isolate soluble target proteins. Furthermore, some foreign proteins are more stable or less toxic when expressed as a fusion protein. One fusion protein expression system that features relatively easy affinity purification of expressed proteins is the pGEX system (8). Maps of three pGEX expression plasmids are shown in *Figure 2*. The pGEX-1 vector is available from AMRAD, while pGEX-2T and pGEX-3X vectors are available from Pharmacia. These pGEX vectors contain the glutathione S-transferase (GST) gene fused at its carboxyl terminus to unique *Bam*HI, *Sma*I, and *Eco*RI sites. These sites are shifted between the vectors, such that fusion with any of the three possible reading frames is possible. A number of derivative pGEX vectors, containing expanded cloning cassettes are also available from Pharmacia. The selectable marker is ampicillin resistance. High level fusion protein expression is driven from an inducible *tac* promoter. The incorporation of the *lacI*q allele that over-produces the lac repressor into the plasmid allows controlled expression in most commonly used *E. coli* hosts, although rearrangements of plasmid DNA have been reported in DH5 derivative or XL-1 blue strains (Pharmacia GST troubleshooting guide). The *tac* promoter is induced in culture by the addition of IPTG.

The pGEX system has several advantages over many other fusion protein expression systems for producing fusion proteins for antibody production. The *tac* promoter is highly inducible, and directs the expression of high levels of fusion protein. The presence of the *lacI*q gene on the plasmid confers relatively tight control of the *tac* promoter and GST expression; this gives broad host specificity and helps establish stocks of plasmids that may encode relatively toxic proteins. The GST protein is rather small (26 kd) compared to other fusion protein systems (e.g. 116 kd for β-galactosidase fusions), and does not interfere with antibody–antigen interactions as do protein A fusions. If necessary, when using the pGEX-2T or pGEX-3X vectors the GST carrier

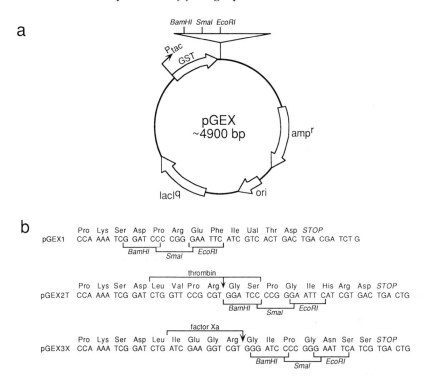

Figure 2. pGEX vectors. (a) Map of a canonical pGEX vector, demonstrating salient details of vector design; ori, amp^r, see *Figure 1* legend; *lacI*^{q}, allele that over-produces the lac repressor; P$_{tac}$, inducible *tac* promoter; GST, glutathione S-transferase gene. (b) Maps of the cloning sites of pGEX1, 2T, and 3X. The reading frame of the three cloning sites (*Bam*HI, *Sma*I, and *Eco*RI) is shifted between the three vectors. The recognition sequences and cleavage sites (shown by *arrows*) are indicated for thrombin and factor Xa.

may also be removed after purification of the recombinant protein by proteolytic cleavage with thrombin or protease factor Xa (see *Figure 2*). GST fusion proteins are often soluble, and are easily purifiable from *E. coli* cell lysates on immobilized glutathione columns. The mild elution conditions allow recovery of relatively pure protein with high antigenicity. However, many fusion proteins are not soluble, and are often difficult to solubilize for affinity chromatography. More seriously, some fusions disrupt the ability of the GST protein to bind glutathione, precluding easy isolation of fusion protein. For this purpose, we suggest the pMAL fusion protein vectors as an alternative system to express foreign genes.

2.6 Fusions with maltose binding proteins

Maps of the pMAL-c, pMAL-cR1, and pMAL-c2 expression constructs (marketed by New England Biolabs) are shown in *Figure 3*. All three vectors

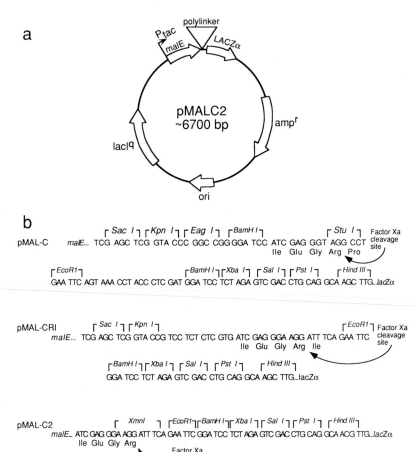

Figure 3. pMAL vectors. (a) Plasmid map of representative pMAL vector, demonstrating the location of the cloning cassette (polylinker) between the *mal*E and *lacZ*α genes; see *Figures 1* and *2* legends for definitions of P$_{tac}$, ori, amp^r, lacI^q. (b) Cloning sites of pMAL-C, -CR1, and -C2. The reading frame of the *mal*E gene is identified in each vector. The cleavage site for protease factor Xa is also shown in each case.

contain a cloning cassette located downstream of the *mal*E gene that facilitates cloning and expression of foreign DNA sequences as fusions with the *mal*E encoded maltose binding protein (MBP; 12, 13). The parent vector has the *mal*E and *lacZ*α genes fused, such that DNA inserts disrupt alpha complementation. This allows individual colonies to be screened for insert by duplicate streaking on ampicillin plates with (tester plate) or without (master plate) IPTG and X-Gal; putative recombinant clones do not form blue colonies on the tester plate. This selection is a significant advantage when large numbers of colonies need to be screened to isolate a difficult recombinant clone. Like the pGEX vectors (*Figure 2*), fusion protein expression is driven by an IPTG inducible *tac* promoter, and the *lacI* gene is present on the plasmid, to reduce expression from the *tac* promoter in the absence of induction. The pMAL-c2 vector uses the *lacI*q allele to further reduce uninduced expression from the *tac* promoter. The high affinity of MBP for maltose allows fusion proteins to be easily purified by, and mildly eluted from, amylose resin columns. However, the ability of the MBP to bind maltose is disrupted in some fusion proteins, especially when the fusions start at restriction sites immediately adjacent to the MBP. This problem is less severe in the pMAL-c2 vector, where a polyasparagine spacer region is inserted between the MBP and the cloning cassette. The MBP is relatively small (42 kd), and if necessary may be proteolytically cleaved from the purified recombinant protein using factor Xa (see *Figure 3*). This series of vectors is suggested as an equally viable alternative to the GST system, if an affinity tagged fusion vector system is required.

2.7 pRIT vectors

Two pRIT vectors, pRIT2T and pRIT5, are available (Pharmacia). In each, the foreign gene sequence is cloned into the C-terminal end of a truncated protein A gene. The resulting expression product can thus be purified by the ability of the protein A fusion partner to bind to IgG–Sepharose. Additional details are given in Chapter 4, Section 2.2.

A potential drawback of these vectors is that protein A fusions may potentially interfere with antibody–antigen interactions. Furthermore, the lambda Pr promoter used to induce protein expression is under lambda repressor control, rather than the more easily inducible *lac* control used with pET, pGEX, and pMAL vectors. The cloning cassettes of these vectors is also not as versatile as that obtainable with the pET, pMAL, or pGEX systems. Due to these disadvantages, we do not recommend these vectors as a first choice for a fusion partner, but they may serve as alternative vectors if needed.

2.8 Pinpoint Xa vectors

These three vectors (three-reading frames; Promega) offer an extensive cloning cassette in which the protein of interest is fused downstream of a sequence that becomes biotinylated during expression in *E. coli*. The biotinylated fusion protein is purified and eluted from a soft release resin (Promega). Although this system offers several potential advantages as a fusion protein expression system, it has only recently become commercially available and hence we are unable to comment on its use based on experience.

3. Construction and screening of expression constructs

3.1 Expression strategies

The general strategy we take, which will be detailed below, is to express 20–60 kd of the protein of interest (which may comprise one or two fragments or the whole protein) in the conveniently designed pET vectors and purify the protein that is most abundantly expressed in inclusion bodies. Whenever a new expression construct is made, it is difficult to predict whether the foreign protein will be toxic or degraded in the cell, or when efficiently expressed, whether the protein will be soluble or insoluble. This section is designed to address these issues and provide a productive, as well as time- and labour-efficient expression strategy for a gene of interest.

3.1.1 Starting materials

We will assume that the starting material is a cDNA clone that has been sequenced, and the open reading frame (ORF) of the encoded protein identified. Computer printouts of restriction maps of this sequence will identify all potentially useful restriction enzyme sites that could be used to clone the ORF, in proper orientation and reading frame, into one of the described expression vectors. Even if useful restriction sites are not identified, the protein coding region can usually be easily cloned via the polymerase chain reaction (PCR) using primers with selected restriction sites at their 5′ ends to amplify, and add these restriction sites to, the relevant region of the gene. Sequence analysis also reveals the predicted molecular weight of the encoded protein, important information that will help identify whether the expressed protein is partially proteolytically degraded. It is possible to express proteins using cDNAs that have not been sequenced, by making fusion clones, in all three reading frames, from identified internal or flanking restriction sites. However, such analyses often fail, since the chosen restriction sites may not be within the ORF or are in the wrong orientation. In addition, the correct product may not be identified due to proteolytic degradation, or, in a GC-rich region, overlapping long ORFs may exist. These problems have been described in detail in a previous chapter in this series (5). Given that modern technology has made the dissection and sequencing of DNA a trivial exercise, it is far more efficient to sequence the cDNA (or at least determine the start of the ORF) *before* designing expression strategies.

3.1.2 Overview of expression strategies

Devising an expression strategy is really a two part process of first determining how much of the protein coding region should be expressed, and second which vector to use. The first question addresses how to define a region of the protein that can be expressed in *E. coli* without major problems. This is often difficult to predict. However, it is clear that the probability of successful expression decreases with the *size* of the expressed protein. In general, expression of fusion proteins is most often successful with proteins less than 60 kd. Very large proteins (> 100 kd) are rarely efficiently expressed, and it is generally more productive to express several smaller fragments of the protein. Furthermore, certain protein regions are, for whatever reason, toxic when expressed in *E. coli*. For example, stretches of foreign DNA containing multiple codons infrequently utilized in *E. coli* are often difficult to express (see Chapter 3, Section 2.1). In the case of toxicity, often the best strategy is to express different fragments of the protein until a construct is found that has eliminated the toxic region and is expressed successfully. Thus one general strategy is to express different modest-sized pieces (40–60 kd) of large proteins and utilize the largest construct that gives the best level of protein expression. In smaller proteins, the initial construct will often express the entire ORF. If this construct is toxic, then smaller derivatives are made in an attempt to eliminate the toxic region. In this manner, moderate-sized portions of even the most troublesome proteins may be successfully expressed.

Given that a fragment of a protein can be successfully expressed, what system should be used to express it? This is primarily guesswork, and a lot depends on whether the expressed protein is soluble or insoluble which cannot be known for certain until the fusion is made. In general, native proteins (as expressed by the pET system) tend to be insoluble, while fusions with the GST or MBP carrier proteins are more often soluble. Protein immunogenicity and antigenicity (as opposed to biological activity) are retained in proteins purified from either insoluble inclusion bodies or from soluble cell lysates. A simple and reliable protocol for isolating relatively pure, immunologically active protein from insoluble inclusion bodies has been devised, and is presented in Section 4.3.1. Thus, we suggest that the initial expression strategy should utilize the pET system to express native proteins. Once part or all of the protein has been successfully expressed, if it is insoluble, the protein may be rapidly extracted from inclusion bodies for use in immunizations. However, if the expressed protein is soluble, then expression of the same DNA fragment in a vector that allows simple affinity purification of soluble protein (e.g. pMAL or pGEX vectors or pET-16b) will facilitate rapid purification of soluble protein. If possible, initial cloning in a poly-histidine tagged pET vector would provide an available affinity tag if the induced protein is soluble. Given the relative ease of inclusion body purification over affinity chromatography, we prefer to purify recombinant proteins from inclusion bodies when possible. This is essentially the opposite strategy employed for isolating biologically active proteins (see Chapter 3), but it is much simpler and generates large amounts of relatively pure antigen for immunization.

3.1.3 Typical expression strategy

An example of our general strategy is shown in *Figures 4* and *5*. In this case, as shown in *Figure 4*, the target protein is the human GMP-140 transmembrane protein. This protein contains several identified peptide domains, including a signal peptide, lectin domain, EGF domain, nine internal protein repeats, a transmembrane domain, and a cytoplasmic domain (14). Two overlapping regions were selected for expression, and these fragments were cloned into both pET and pGEX expression vectors, and expressed (*Figure 5A*). The

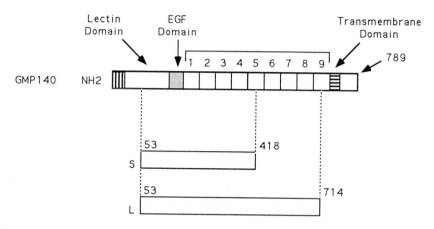

Figure 4. Overview of the human GMP-140 protein. The structural domains of the protein are shown. These are the N-terminal signal peptide (vertically striped region), lectin domain, EGF domain, and transmembrane domain, as well as nine numbered internal repeats. Two regions, either small (S) or large (L), that were cloned in both pET and pGEX vectors are shown *below*. Numbers in the S and L regions denote the amino acid endpoints contained in these fragments.

larger construct is not efficiently expressed in the pET vector. Fusion protein is produced from the pGEX vector but it appears to be partially degraded. However, the smaller construct is efficiently expressed in both systems; clearly more protein is made with the pET construct, probably due to fact that the T7 promoter is stronger than the *tac* promoter (*Figure 5A*). Both induced proteins were shown to be insoluble (*Figure 5B*). Due to the higher expression level, and lack of carrier protein sequences, the pET expression construct was used in a large scale preparation to purify large quantities of the insoluble, relatively pure, inclusion body for immunization purposes.

3.2 Construction and screening of pET expression constructs

Foreign DNA is inserted into restriction sites of an appropriate pET vector, utilizing naturally occurring restriction sites, or synthetic linkers to add appropriate restriction sites, or by adding restriction sites during PCR amplification. If possible, direct the orientation of the insert by generating two different cohesive ends. A brief overview of required cloning methodology will be given here. However, detailed descriptions of these techniques are available elsewhere (15).

3.2.1 Preparation of expression vector DNA

Given that pET vectors do not allow selection for insert, it is important to limit the amount of intramolecular ligation of vector and thus reduce the

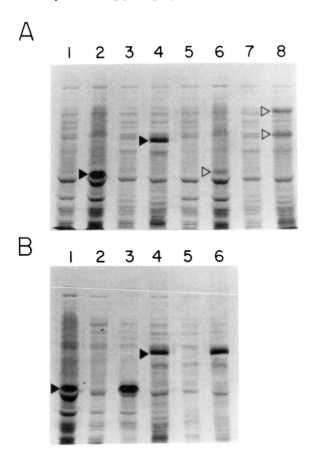

Figure 5. Expression of fragments of the GMP-140 protein in pET and pGEX vectors. (A) SDS–PAGE analysis of protein expression from various GMP-140 expression constructs. Total protein preparations from uninduced and induced cultures containing GMP-140 fusion constructs were analysed by SDS–PAGE and stained with Coomassie blue dye (*Protocol 3*). All inductions were for three hours. The constructs express either the small (lanes 1–4) or large (lanes 5–8) DNA fragment of *Figure 4*, in either pET-3c (lanes 1, 2, 5, and 6) or pGEX-2T (lanes 3, 4, 7, and 8). *Filled arrowheads* indicate protein induced from the pET and pGEX constructs containing the smaller GMP-140 fragment. Note that, although more degradation products are formed (*open arrowheads*), expressing the large fragment as a gene fusion in pGEX significantly increases protein stability as compared with the pET construct. (B) Solubility of induced proteins was determined as described in *Protocol 4*, for the pET (lanes 1–3) and pGEX (lanes 4–6) vectors expressing the small GMP-140 fragment. *Arrows* denote the induced protein present in the total protein lanes (lanes 1 and 4), for comparison with the soluble (lanes 2 and 5) or insoluble (lanes 3 and 6) protein samples. Note that neither induced protein is present in the soluble fractions, demonstrating that both proteins are insoluble.

number of non-recombinant transformants that are recovered. This may be accomplished in two ways. Preferably, the cloning will be directional, and require cutting the vector with two non-compatible restriction enzymes (e.g. *Nde*I and *Bam*HI in the pET-3 series). Gel purification of the doubly-cut vector will remove the intervening linker region, and background non-recombinant clones will be greatly reduced. Alternatively, if the vector is singly cut, treating the vector with calf intestinal phosphatase (CIP) will remove 5' phosphates from the ends of the vector DNA, preventing intramolecular ligation. *Protocol 1* describes these procedures.

Protocol 1. Preparation of vector DNA

Equipment and reagents

- Standard reagents for plasmid DNA purification either by CsCl centrifugation or polyethylene glycol precipitation (15), or a commercially available plasmid purification kit (e.g. Qiagen Midi or Maxi columns)
- Appropriate restriction enzymes for vector cleavage
- Calf intestinal phosphatase (CIP) (optional)
- Equilibrated phenol/chloroform (50:50 v/v). Prepare this as described in ref. 15

- TE buffer: 10 mM Tris–HCl 8.0, 1 mM EDTA
- 0.8% agarose gel containing 0.25 μg/ml ethidium bromide plus other reagents required for gel electrophoresis
- Standard DNA loading buffer: e.g. 0.25% bromophenol blue, 0.25% xylene cyanol, 30% glycerol in H_2O (store at 4 °C)
- 3 M sodium acetate pH 5.5

Method

1. Purify the vector DNA by standard methodology (15). Although CsCl banded DNA is optimal, purification of DNA from 50 ml cultures utilizing polyethlylene glycol precipitation (15) or a commercially available kit (e.g. Qiagen) is quicker, and generates DNA sufficiently pure for subsequent manipulations. If CIP treatment is anticipated, it is important to ensure that the plasmid is RNA-free, since RNA will compete as a substrate for alkaline phosphatase.

2. Cleave 1–5 μg of the DNA in the cloning site with the necessary restriction enzyme(s), generally at a concentration of 100 μg DNA/ml.

3. Extract the sample once with phenol/chloroform and once with chloroform, centrifuging 2 min in a microcentrifuge after each extraction (15).

4. Ethanol precipitate the DNA by adding 0.1 vol. 3 M sodium acetate pH 5.5 and 2 vol. of 99% ethanol. Mix, and then store at −20 °C for more than 10 min.

5. Spin for 10 min in a microcentrifuge at 4 °C.

6. Wash the pellet with 1 ml 70% ethanol, dry, and resuspend it at 1 μg/ 10 μl in sterile TE buffer (use H_2O if the sample is to be CIP treated).

7. (Optional) If desired, treat the DNA sample with CIP using standard methodology (15). After CIP treatment, phenol/chloroform extract, ethanol precipitate, and resuspend DNA in TE buffer as described in steps 3–6.

8. Analyse the integrity of the DNA (0.5–1 μg) by electrophoresis in a 0.8% agarose gel.

9. If the vector is not treated with CIP, isolate the resolved vector DNA from the gel, for example utilizing the Prep-a-Gene matrix (Bio-Rad).

10. Store the purified vector DNA at −20°C.

3.2.2 Preparation of foreign DNA insert

Approximately 1–5 μg of foreign DNA is cut with the appropriate restriction enzyme(s). After restriction, the DNA end(s) must be blunt if DNA linkers are to be added. This is accomplished utilizing T4 polymerase to remove extending 3′ ends or the Klenow fragment of DNA polymerase to fill recessed 3′ ends (15). Linkers are ligated to the blunt-ended DNA, and cohesive ends are generated by restriction digestion with the appropriate enzyme. This restriction digestion step is not necessary if DNA adaptors are used instead of linkers. If linkers are added to only one end of the fragment, it may be necessary to restrict the second end of the insert after the linker ligation, to prevent the disruption of this second cohesive end during the blunt-ending or linker ligation steps. The prepared foreign insert DNA is resolved and purified on agarose gels as described for vector DNA in *Protocol 1*.

Recently, an elegant alternative to standard restriction enzyme cloning has become available. In this method, PCR is utilized to amplify and clone any DNA fragment. A brief overview of this procedure, as pertains to cloning in a pET-3-based expression vector is diagrammed in *Figure 6*. Often, an expression construct can be made that contains no fusion protein sequences and expresses the entire target ORF (see *Figure 6*). This technology is fast and convenient, and is particularly useful in cases where useful restriction sites are not present in the regions of interest. It is important to note that PCR amplification with *Taq* polymerase is not recommended, due to the high error rate of this enzyme. The proofreading thermostable *Pfu* (Stratagene) or ULTMA (Perkin Elmer) polymerases work well for this purpose. A detailed description of this protocol, including PCR amplification conditions and primer designing tips, is available elsewhere (16).

3.2.3 Ligation, transformation, and characterization of pET fusions

Purified vector DNA and foreign DNA are ligated together and used to transform *E. coli* competent cells, plating the cells on LB plates containing

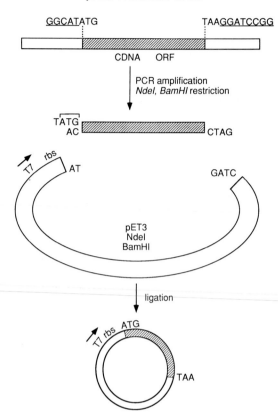

Figure 6. Schematic representation of PCR cloning into a pET-3 expression vector. A primer is designed starting at the ATG start codon and including another 15–18 bases 3' to the start codon (not shown). Additional bases are added to the 5' end, such that the ATG site is converted into a *Nde*I restriction site (*underlined*; an alternative primer site is selected if the amplified region contains an internal *Nde*I site). A second primer is con-structed at the 3' end of the fragment, again including 15–18 bases of homology, followed by additional bases that create a *Bam*HI site (*underlined*), or a compatible *Bgl*II or *Bcl*I site if a *Bam*HI site is present within the insert. PCR amplification is performed using these two primers, and the selected region is both amplified and modified to contain asymmetric *Nde*I and *Bam*HI sites. Restriction digestion with *Nde*I and *Bam*HI allows this fragment to be purified and directionally cloned into similarly cut pET-3 vector, creating an expression construct that contains no fusion protein sequences.

ampicillin (*Protocol 2*). Ampicillin-resistant colonies are re-streaked on to fresh plates. If the frequency of insert is expected to be low, then putative recombinants may be selected based on size, from rapid screens of disrupted bacterial colonies (*Protocol 2*). In this manner, recombinant clones can be easily identified, even if represented at a low frequency (< 5%) in the recovered ampicillin-resistant colonies. If cloning in pMAL vectors, duplicate streaking of colonies on LB plates containing either ampicillin (master plate) or ampicillin, 50 µg/ml X-Gal, and 0.1 mM IPTG (tester plate) identifies putative recombinant clones by lack of blue staining on the tester plate. Plasmid DNA from putative recombinants is isolated from 2 ml 2×YT (containing 100 µg/ml ampicillin) cultures by the alkaline lysis method (15), and the presence and orientation of insert DNA determined by restriction analysis. Once a clone with the desired restriction map is identified, sequence analysis of the T7 promoter, ribosome binding site, and recombinant clone junction, utilizing a commercially available primer (New England Biolabs), definitively identifies whether the structure of the fusion is as predicted.

Protocol 2. Rapid screening for recombinant clones

Equipment and reagents

- Purified vector DNA (100 ng/µl) from *Protocol 1*
- Purified foreign DNA (200–500 ng/8 µl TE buffer)
- 10 × ligation buffer: 200 mM Tris–HCl pH 7.6, 50 mM MgCl$_2$, 10 mM DTT, 500 µg/ml nuclease-free BSA, 10 mM rATP (store at −20 °C up to one month)
- T4 DNA ligase
- Competent cells for transformation (e.g. JM109)

- LB plates, containing 50–100 µg/ml ampicillin
- Cell disruption buffer: 10 mM Tris–HCl pH 8.0, 100 mM NaCl, 10 mM EDTA
- Phenol/chloroform mix (*Protocol 1*)
- RNase A stock (1 mg/ml in TE buffer)
- DNA loading buffer (*Protocol 1*)
- 0.8% agarose gel (*Protocol 1*)

Method

1. For ligation, mix in a microcentrifuge tube:

   - purified vector DNA (100 ng)   1 µl
   - gel purified foreign DNA[a] (200–500 ng)   8 µl
   - 10 × ligation buffer   1 µl
   - T4 DNA ligase (1–2 U)   0.5 µl

   Incubate for 4 h to overnight at 16 °C.

2. Use 3–4 µl of the ligation mixture (from step 1) to transform competent *E. coli* cells by standard methodology (15).

3. Plate the transformed cells on LB plates containing 50–100 µg/ml ampicillin.

4. Make short streaks of potential recombinant colonies from transformation plates on to fresh LB ampicillin plates (multiple streaks per plate) and grow overnight or until significant bacterial growth is obtained.

5. Resuspend a small amount of the streak in 40 µl of cell disruption buffer in a 1.5 ml microcentrifuge tube. Add 50 µl phenol/chloroform, and vortex the tube for 1 min.

6. Centrifuge for 1 min at high speed in a microcentrifuge.

Protocol 2. *Continued*

7. Add the supernatant to 1 μl of RNase A stock solution in a fresh tube, and incubate for 10 min at room temperature.

8. Add 10 μl of DNA loading buffer and heat the sample for 5 min at 70 °C. Carefully load 15 μl of the DNA on to a 0.8% agarose gel (the DNA may be very viscous) and electrophorese.

9. Photograph the gel. Plasmids containing the insert are identified by comparison of plasmid mobility with that of similarly prepared parent vector DNA.

---

[a] The insert DNA should be at one to five molar excess.

3.2.4 Screening for T7-driven protein induction

Once a confirmed pET fusion construct has been obtained, the plasmid DNA must be transformed into a DE3 lysogen host to induce foreign protein expression. High efficiency competent cells are not needed, since transformation utilizes purified DNA clones. Indeed, stocks of competent cells from the BL21(DE3), BL21(DE3)LysS, and BL21(DE3)LysE strains may be made and stored for several years at −70 °C. Generally constructs are initially transformed into only the BL21(DE3) and BL21(DE3)LysS strains, since LysE stocks show reduced protein induction, and only very toxic foreign proteins will require expression in the LysE strain. Indeed, the low level of T7 lysozyme produced by the LysS plasmid aids during the preparation of cell extracts. After cell preparations are freeze-thawed to disrupt the cell membrane, the T7 lysozyme is released from the cell interior, and cleaves a bond in the peptidoglycan layer of the *E. coli* cell wall, leading to cell lysis (9). Since the presence of the LysS plasmid is neither detrimental nor reduces fusion protein levels, often expression constructs are transformed only into BL21(DE3)LysS for preliminary protein induction. Protein induction is tested as described in *Protocol 3* (see *Figure 5A* for sample induction). If protein expression is successful, the solubility of the induced protein is determined, as described in *Protocol 4*, and shown in *Figure 5B*. An example of the influence of host strains on the level of protein induction is shown in *Figure 7* (lanes 1–3) for the *fushi tarazu* protein of *Drosophila*.

Protocol 3. Small scale induction of protein expression

Equipment and reagents

- LB plates, containing 50–100 μg/ml ampicillin
- Choramphenicol stock solution (34 mg/ml in ethanol)
- Ampicillin stock solution (50 mg/ml in H₂O)
- LB medium
- 2 × YT medium: 16 g tryptone, 10 g yeast extract, 5 g NaCl per litre, pH 7.5
- IPTG stock solution (0.2–1.0 M in H₂O)

- 2 × sample buffer: 0.125 mM Tris–HCl pH 6.8, 2 mM EDTA, 6% SDS, 20% glycerol, 0.025% bromophenol blue. β-mercaptoethanol is added to 5% before use
- Protein molecular weight standards (e.g. Sigma or Bio-Rad)
- 7.5% or 12% SDS–polyacrylamide gel (depending on the size of the induced protein)

Method

1. Transform the plasmid DNA into competent expression host cells and plate on LB ampicillin plates. If using LysS or LysE containing strains, choramphenicol may be added to the plates by spreading 100 μl of dilute chloramphenicol (15 μl stock solution in 985 μl ethanol) on each ampicillin plate. Incubate overnight at 37 °C and then re-streak a single colony.

2. Inoculate 5 ml of LB media (containing 100–200 μg/ml ampicillin, including 34 μg/ml chloramphenicol for LysS and LysE cells) with a single colony. Let the tube *stand* overnight at 37 °C. Alternatively, streak a fresh antibiotic plate with a single colony and incubate overnight at 37 °C.

3. Inoculate 10 ml of 2 × YT (containing appropriate antibiotics) with either 50 μl of the overnight culture (take care to ensure that the culture is not grown to saturation) or with bacteria scraped from the fresh plate. Incubate with aeration at 37 °C until the culture reaches 0.4–0.6 OD_{600}.

4. Remove a 1 ml sample to a 1.5 ml microcentrifuge tube and centrifuge for 1 min. Discard the supernatant and resuspend the pellet in 150 μl of 2 × sample buffer. This is the uninduced protein sample which may be stored at −20 °C. Immediately induce the remaining culture by adding IPTG to a final concentration of 0.4–1 mM and resume incubation.

5. After 1–3 h, remove a sample and process it as in step 4. If performing a time course optimization, remove and process samples at several intervals after induction.

6. Heat all samples to 95 °C for 5 min and clarify by centrifugation for 1 min in a microcentrifuge. Load 10 μl of each sample on to an SDS–polyacrylamide gel. Use a 7.5% gel unless the induced protein is expected to be < 30 kd in which case a 12% gel should be used. Apply protein molecular weight standards in adjoining lanes. Electrophorese until the bromophenol blue dye migrates to the end of the gel.

7. Fix and stain the gel with Coomassie blue dye after electrophoresis. Induced proteins are identified by comparison with the uninduced protein control lane.

Protocol 4. Determination of protein solubility

Reagents

- Reagents for growth and induction of culture and preparation of induced protein (*Protocol 3*)
- Lysis buffer: 50 mM Tris–HCl pH 8.0, 2 mM

EDTA, 100 μg/ml lysozyme (added immediately before use)
- 2 × sample buffer (*Protocol 3*)

Method

1. Grow and induce a 10 ml culture and prepare a sample of induced protein as in *Protocol 2*. This is the total protein sample.

2. Centrifuge a second 1 ml sample in a microcentrifuge tube for 20 sec and resuspend in 150 μl of lysis buffer. Incubate for 15 min at 30 °C.

3. Sonicate with a microtip, two times for 10 sec at 4 °C.

4. Centrifuge for 2 min in a microcentrifuge and remove the supernatant to a fresh tube containing an equal volume of 2× sample buffer. This is the soluble protein extract. Resuspend the pellet in 150 μl 2 × sample buffer for a insoluble protein extract.

5. Heat, and analyse 10 μl aliquots of the samples by SDS–PAGE as in *Protocol 3*. Although the pellets are contaminated with soluble proteins, absence of the induced protein in the soluble fraction is a reliable indicator of insolubility.

3.3 Cloning and expression in pGEX and pMAL vectors

Although the cloning sites in pGEX and pMAL vectors are different from the pET vectors, and thus restriction enzymes used during cloning will vary, the methodology involved during preparation and ligation of vector and insert DNA is the same as that described for the pET system. However, since the pGEX and pMAL systems utilized endogenous *tac* promoter for expression, the same *E. coli* strains may be used for both plasmid analysis and protein induction. This can be a disadvantage of this system, since the low levels of fusion protein expression that accumulate in these hosts will prevent establishment of some toxic plasmids. Sequence analysis of the pGEX clone junction may utilize a sequencing primer (5′ GCATGGCCTTTGCAGGG 3′) corresponding to sequences upstream of the pGEX cloning cassette (Pharmacia GST troubleshooting guide) while the corresponding region in pMAL fusions is sequenced using a *malE*-specific primer (5′ GGTCGTCAGACT-GTCGATGAAGCC 3′) available from New England Biolabs. Once pGEX or pMAL recombinants have been characterized, protein expression and solubility tests are as for the pET system (*Protocols 3* and *4*).

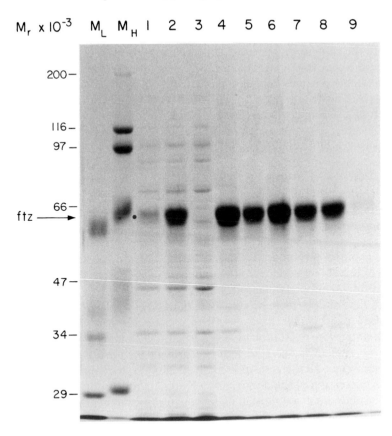

Figure 7. Optimization of inclusion body induction and purification. SDS–PAGE analysis of the expression of the *fushi tarazu* (ftz) protein (*arrow*) in *E. coli* strains BL21 (DE3), lane 1; BL21 (DE3) pLysS, lane 2; and BL21 (DE3) pLysE, lane 3. The highest level of induction is in the pLysS strain which was used to investigate means of purifying inclusion bodies away from host proteins. Inclusion bodies were washed with: lane 4, 50 mM Tris–HCl pH 8.0, 150 mM NaCl, 1 mM EDTA; lane 5, 0.1% SDS; lane 6, 0.1% SDS, 1% Triton X-100, 1% sodium deoxycholate; lane 7, 0.5 M guanidine-HCl; lane 8, 1.0 M guanidine-HCl; and lane 9, 2.0 M guanidine–HCl. The purest preparation is in lane 6. M_L and M_H are low and high protein molecular weight marker ladders (Sigma).

3.4 Troubleshooting and optimization

If protein expression is induced from the expression construct, then the expression procedure is optimized for subsequent large scale protein induction. In general, an examination of the time course of protein induction is necessary to determine the optimal time after induction for cell harvesting, generally between a half and three hours after induction. Protein integrity must also be determined (see below). A large scale culture is then grown (see Sections 4.1 and 4.2) and, if the induced protein is insoluble, protein is purified from the inclusion bodies (Section 4.3.1). However, if the induced protein is soluble and in a non-tagged pET vector, it must be decided whether to purify the protein by preparative gel electrophoresis (Section 4.3.2), or to re-clone the gene into a tagged expression vector that allows affinity purification of soluble protein (Section 4.3.3).

However, protein expression will often initially not be satisfactory, and some troubleshooting will be required. There are two general types of problems that are commonly encountered when expressing proteins in *E. coli*— either absence of fusion protein induction, or degradation of fusion proteins.

3.4.1 Absence of inducible fusion protein

This includes problems generally related to plasmid toxicity. As discussed above, on occasion a recombinant protein will be too toxic to be established in *E. coli* hosts. In other cases establishment is achieved, but no inducible protein is observed. This may be attributable to rapid cell death after foreign protein expression is induced. In these cases, removing protein samples at 15 minute intervals immediately after induction may detect protein induction that is subsequently lost due to cell lysis.

Often lack of detectable protein induction is due to plasmid loss. All of the expression plasmids we have described contain the β-lactamase gene that confers ampicillin resistance to *E. coli* cells containing the vector. The β-lactamase protein is secreted into the growth medium, where it destroys ampicillin. Once all the ampicillin in the medium is destroyed, cells that have lost the expression plasmid may grow. Normally this is not a problem, since most plasmids are maintained at high enough copy number that the plasmid is retained even in the absence of ampicillin selection. However, if the foreign protein contained in the expression plasmid is toxic, the low level of protein produced without IPTG induction will result in impaired cell growth, low plasmid copy number, and strong selection against cells carrying the plasmid. Once enough β-lactamase has been made and secreted into the medium to destroy all the ampicillin, the culture is overgrown by cells that have lost the plasmid. Thus, failure to detect fusion protein expression is often due to the fact that, when induced, most cells of the culture have lost the plasmid.

A few simple precautions can prevent plasmid loss. Functionally, even when high levels of ampicillin are used (200 μg/ml ampicillin) loss of selection will occur before the culture becomes turbid. Furthermore, diluting saturated cultures into fresh media is not an effective means of restoring selection, since saturated cultures contain significant amounts of soluble β-lactamase that will quickly destroy the ampicillin in the fresh culture (9). These facts, plus the observation that many toxic genes that are tolerable in growing cultures kill cells in saturated cultures, indicate that cultures of expression plasmids should *never* be allowed to reach saturation. Functionally, this may be achieved by growing culture inoculums as standing cultures (to reduce aeration and growth rate), and using only cultures that are lightly turbid as inoculum stocks. Alternatively, inocula may be grown on plates. A freshly grown single colony is streaked on to a culture plate containing 100 μg/ml ampicillin and grown overnight, the resulting bacteria are directly inoculated into liquid culture, grown, and induced. In addition, the use of carbenicillin in place of ampicillin has been reported to prevent culture overgrowth by cells that have lost the plasmid (Novagen pET system manual). If, after these precautions, no protein expression is detected, the percentage of cells containing the expression construct at the time of induction may be directly determined (9), as described in *Protocol 5*. If most induced cells are shown to contain the plasmid, then plasmid instability may be ruled out as the problem.

Unfortunately, once plasmid instability is ruled out and sequence analysis of the cloning site has confirmed that the clone has the correct reading frame fusion, very little can be done to obtain sufficient levels of expressed protein for immunization purposes. Although a number of potential reasons for toxicity are known (see ref. 9 for a discussion) this knowledge is not particularly consoling. In terms of general strategy, it is probably most useful to start over again, with smaller or overlapping constructs, in an attempt to identify a fusion that is not toxic.

Protocol 5. Determination of plasmid stability

Equipment and reagents

- Reagents for culture growth and induction (*Protocol 3*)
- LB plate (plate A)
- LB plate containing 50–100 μg/ml ampicillin (plate B)
- LB plate containing 1 mM IPTG (plate C)
- LB ampicillin plate including 1 mM IPTG (plate D)

Method

1. Grow a culture and induce as in *Protocol 3*. Immediately before induction, remove an aliquot of cells for plating analysis.

2. Serially dilute the culture aliquot, and spread aliquots of 100–200 cells on plates A–D. Assume that 1 OD_{600} = 8 × 10^8 cells/ml. However, this number will vary between strains (e.g. this number will be lower for RecA$^-$ cells) and a couple of different dilutions should be plated. Incubate the plates overnight.

3. Count the colony number on all four plates. Comparison of colony numbers on plates A and B determines the percentage of cells containing the plasmid at induction. In a typical experiment, it is not unreasonable to expect that 98% of the cells will contain the plasmid. All cells containing functional plasmid should either not grow [in the case of pET vectors in (DE3) or (DE3)Lys S cells] or form only small colonies (pGEX and pMAL vectors) on IPTG plates. Thus comparison of the numbers of colonies on plates D and B establishes the percentage of plasmids that are functional.

3.4.2 Degradation of fusion proteins

Often fusion protein induction will be observed, but the accumulated protein is either considerably smaller than predicted, or present as multiple discrete bands (see *Figure 5*A). These observations are diagnostic of protein degradation (this may also be one cause of absence of protein induction). The T7 expression hosts are apparently deficient in both the *lon* and *ompT* proteases and are generally limited in protein degradation relative to other *E. coli* K-12 strains. However, protein degradation of unstable fusion proteins varies unpredictably between strains, and it may be useful to assess protein stability in a number of different *E. coli* expression hosts. Protein stability may also be enhanced in certain strains of *E. coli* that are distantly related to K-12 (most expression host strains are derivatives of K-12). Six of these strains, referred to as TOPP strains, are commercially available from Stratagene. Transformation and protein induction in these strains may improve protein stability, allowing protein purification without re-designing the expression construct. If a native protein is degraded, often expression of this protein as a fusion will enhance stability (see *Figure 5* for an example), particularly when the native protein is soluble. In a similar manner, insoluble native protein may accumulate without degradation in inclusion bodies, but may be degraded when expressed as a soluble fusion protein. Given that the properties of a foreign protein may vary when expressed as either a native or a fusion protein, and may be much easier to purify in one form but not the other, it is often valuable to clone and express a gene in both native (e.g. pET) and fusion (e.g. pGEX) forms.

3.5 Verifying protein integrity

Once a fusion protein is successfully induced, it is important to determine if the detected protein corresponds to the full-length protein, or is a degradation product. Although sequence analysis of the foreign DNA insert allows a prediction of the protein molecular weight, in practice the mobility of proteins are often anomalous in polyacrylamide gels. For example, a high proline content will often increase the apparent molecular weight of a protein, such that a major degradation product may migrate anomalously near the predicted M_r of the full-length product (see ref. 5). To reduce the chances of such misleading results, Western blot analysis (*Protocol 6*) of protein gels containing the induced fusion protein is advised. This is recommended because immunological detection is very sensitive, and will detect low levels of a higher molecular weight precursor protein that would be present if degradation has occurred. In the case of the pET vectors, if the leader peptide is retained in the fusion, Western analysis may be performed utilizing an antibody specific to this peptide (available from Novagen). A GST-specific antibody is available from AMRAD for similar tests of pGEX inductions, while pMAL fusions may be detected with an anti-MBP antiserum (New England

Biolabs). Antibodies specific to epitopes expected to be present on the foreign protein will also aid clone characterization.

Protocol 6. Western blot analysis

Equipment and reagents

- Induced protein samples (*Protocol 3*)
- 7.5% or 12% SDS–polyacrylamide gel (*Protocol 3*)
- Protein electroblotting apparatus (e.g. ABN polyblot system)
- Nitrocellulose filters, 0.45 μm pore size (Schleicher and Schuell)
- Ponceau S concentrate: 2% Ponceau S, 30% trichloroacetic acid, 30% sulforalicylic acid (Sigma)
- Phosphate-buffered saline including Tween-20 (PBST): 10 mM $NaPO_4$, 150 mM NaCl, including 0.1% Tween-20, pH 7.0
- Blocking buffer: 5% non-fat dry milk in PBST
- BBS-Tween buffer: 0.1 M boric acid, 0.025 M sodium borate, 1 M NaCl, 0.1% (v/v) Tween-20, pH 8.3
- Primary antibody

- Species-specific, alkaline phosphatase-conjugated secondary antibody (Boehringer-Mannheim)
- Species-specific, biotin-conjugated secondary antibody (optional; Boehringer-Mannheim)
- Streptavidin–alkaline phosphatase conjugate (optional; Boehringer Mannheim)
- Nitroblue tetrazolium stock solution (Sigma); 75 mg/ml nitroblue tetrazolium in dimethyl formamide (store at −20°C)
- 5-Bromo-chloro-indolylphosphate stock solution (Sigma): 50 mg/ml in dimethyl-formamide (store at −20°C)
- Alkaline phosphate substrate buffer: 100 μg/ml nitroblue tetrazolium, 50 μg/ml 5-bromo-chloro-indolylphosphate, 5 mM $MgCl_2$ in 50 mM Na_2CO_3, pH 9.5.

Method

1. Prepare and electrophorese the induced protein samples, along with protein molecular weight standards, on SDS–PAGE as described in *Protocol 3*. Transfer the separated proteins from the polyacrylamide gel to nitrocellulose using an electroblotting apparatus following the manufacturer's instructions.

2. After transfer, temporarily stain the nitrocellulose blots for 3 min with Ponceau S solution diluted 1:10 in distilled H_2O to visualize the bands. Destain the blot by running a gentle stream of distilled H_2O over the blot. The stained blot will indicate the efficiency of transfer and the location of sample lanes and markers. Mark the lane borders and protein standard locations on the blot with a pencil. Rinse the blot in PBST.

3. Block protein binding sites on the filter by immersing the blot in blocking solution overnight at 4°C, or 2 h at room temperature (15 min at room temperature is sufficient if maximal sensitivity is unnecessary).

4. Incubate the blot with the appropriate primary antibody, usually diluted to about 1.0–2.5 μg/ml in blocking solution, for 1–2 h at room temperature.

5. Remove unbound primary antibody by washing the blot with two changes each of large volumes of PBST, BBS-Tween, and PBST successively (10 min/wash). Extensive washing is important to minimize background staining.

Protocol 6. *Continued*

6. Incubate the blot for 1–2 h at room temperature with the species-specific, alkaline phosphatase-conjugated secondary antibody, diluted to about 1 μg/ml in blocking solution.

7. (Optional) Alternatively, for greater sensitivity, the blot may be developed to an alkaline phosphatase conjugate in two steps from the primary antibody. First, a biotin-conjugated secondary antibody is used as indicated in step 6, then washed as in step 5. The blot is then incubated with streptavidin-conjugated alkaline phosphatase as in step 6. This amplification of signal should be unnecessary to detect low level expression of most fusion protein.

8. Wash the blot as in step 5, substituting 50 mM Na_2CO_3 pH 9.5 for PBST in the final wash.

9. Develop the blot in freshly prepared alkaline phosphatase substrate buffer. Development, which may take from a few minutes to several hours, is stopped by flooding the blot with distilled H_2O, and the blot is subsequently air dried and stored. The blot will fade after several hours exposure to light.

4. Expression and purification of recombinant antigens

The goal of the protocols in this section is to obtain antigen of sufficient quantity and purity for immunizing animals. With large scale cultures, obtaining milligram (at a minimum) quantities of antigen should be possible, although depending upon the recombinant protein, the final yield can vary tenfold or more. In addition to yield, the purity and integrity of the final protein suspension will also be antigen-specific. Each of these variables can be easily assayed by SDS–PAGE. In general, protein yield is the most critical parameter for use in antibody production and purification. These protocols are generally not suitable for obtaining biologically active proteins (e.g. active transcription factors); however, biological activity is not necessary to elicit an immune response and obtain high titre antisera.

4.1 Induction of protein expression

Once optimal induction conditions (e.g. vector, host strain, induction time) are known for an individual recombinant protein (see Section 3.4), large scale cultures ($\geq$ 500 ml) are grown and induced (*Protocol 7*). The best yield usually results from keeping a fresh culture (i.e. inoculating with a colony from a fresh plate, and avoiding letting the culture reach stationary phase prior to induction). Rich media such as 2×YT give better protein yields although such complex media may contain enough inducer-like molecules to prevent growth of strains expressing some toxic proteins. The culture is grown to midlog phase and induced by adding IPTG to a final concentration of 0.4–1.0 mM. The culture is then grown for the length of time determined to be optimal from a time course experiment (Section 3.4), generally at least two hours. The level of protein induction in the culture can be assayed by SDS–PAGE (*Protocol 3*), comparing a sample taken just prior to induction with a sample taken after full induction (see *Figure 9*, lanes 1 and 2 for an example of an appropriate level of induction).

Protocol 7. Large scale induction of protein expression

Equipment and reagents

- LB plates containing 100 μg/ml ampicillin (+ 34 μg/ml chloramphenicol if strain is pLysS)
- 500 ml 2×YT medium (*Protocol 3*) containing 100 μg/ml ampicillin (include 10 ml of 20% glucose if growing a pMAL construct to prevent induction and co-purification of the *E. coli* maltose binding protein)
- 34 mg/ml chloramphenicol (in ethanol)
- IPTG stock solution (*Protocol 3*)
- 2 × sample buffer (*Protocol 3*)
- Spectrophotometer
- 7.5% or 12% SDS–polyacrylamide gel (*Protocol 3*)
- Other reagents; see *Protocol 2*

Method

1. Streak a fresh colony of the appropriate strain on to an LB plate containing 100 μg/ml ampicillin and grow overnight at 37 °C.

2. Pick one colony and inoculate a fresh culture of 5 ml 2×YT containing 100 μg/ml ampicillin (± chloramphenicol) and incubate overnight at 37 °C *without* rotating (i.e. a standing overnight culture). Alternatively, streak a fresh medium plate and incubate overnight at 37 °C.

3. Use the entire 5 ml (or scrape most of the cells off the plate) to inoculate 500 ml 2×YT containing 100 μg/ml ampicillin and incubate with shaking until $OD_{600} = 0.5$.

4. Remove 1 ml of the culture (the uninduced sample) to a 1.5 ml microcentrifuge tube and centrifuge for 1 min. Discard the supernatant and resuspend the pellet in 150 μl of 2 × sample buffer.

5. Add IPTG to a final concentration of 0.4 mM.

6. Grow for 2 h at 37 °C with shaking (the length of induction depends upon the previously optimized time).

7. Remove 1 ml of the culture (the induced sample) to a 1.5 ml microcentrifuge tube and centrifuge for 1 min. Discard the supernatant and resuspend the pellet in 150 μl of 2 × sample buffer.

8. Heat the uninduced and induced samples for 5 min at 95 °C and clarify by centrifugation for 1 min in a microcentrifuge.

Protocol 7. *Continued*

9. Analyse 10 μl by SDS–PAGE using a 7.5% or 12% polyacrylamide gel (see *Protocol 3*) and stain with Coomassie blue to assay the level of protein induction.

4.2 Cell harvesting and lysis

After adequate induction of protein expression, cells are harvested by centrifugation and lysed (*Protocol 8*). Lysis and subsequent release of protein is accomplished by a combination of lysozyme treatment, freeze-thaw, and ultrasonication. The culture is kept chilled and in the presence of protease inhibitors to reduce proteolysis. When lysis is complete, the recombinant protein can be purified by one of the procedures described in Section 4.3.

Protocol 8. Cell harvesting and extraction

Equipment and reagents

- Ultrasonicator
- Inclusion body sonication buffer: 25 mM Hepes pH 7.7, 100 mM KCl, 12.5 mM MgCl$_2$, 20% glycerol, 0.1% (v/v) Nonidet P-40, 1 mM DTT
- Nickel tag sonication buffer: 20 mM Tris–HCl pH 7.9, 500 mM NaCl, 5 mM imidazole
- pMAL sonication buffer: 20 mM Tris–HCl pH 7.4, 200 mM NaCl, 10 mM EDTA, 10 mM EGTA, 10 mM β-mercaptoethanol

- pGEX sonication buffer (PBS)
- 10 mg/ml lysozyme
- 100 mM PMSF in absolute ethanol or isopropanol; store at −20 °C
- 10 mg/ml leupeptin (Boehringer) in H$_2$O; store at −20 °C
- 1 mg/ml pepstatin (Boehringer) in methanol; store at −20 °C
- 10 mg/ml aprotinin (Boehringer) in H$_2$O; store at −20 °C

Method

1. Transfer the induced culture to two 250 ml centrifuge bottles and chill the cells on ice for 10–15 min. Most protein degradation occurs during cell harvesting and lysis. Therefore perform these steps quickly and keep the cells chilled.

2. Centrifuge the cells at 3000 *g* for 10 min at 4 °C to pellet the cells.

3. Resuspend the cell pellet thoroughly in 10 ml of the appropriate sonication buffer. Transfer to a 30 ml non-glass centrifuge tube.

4. Add 100 μl 100 mM PMSF, 2 μl 10 mg/ml leupeptin, 20 μl 1 mg/ml pepstatin, and 4 μl 10 mg/ml aprotinin to minimize proteolysis.

5. Add 500 μl of 10 mg/ml lysozyme and incubate on ice for 30 min (the culture should become quite viscous).

6. Freeze the suspension at −70 °C for at least 1 h (the suspension may be left at this temperature indefinitely).

7. Thaw the suspension rapidly in a water-bath at room temperature and then place on ice.

8. Sonicate the suspension *thoroughly* using at least eight 15 sec bursts of the microprobe with intermittent cooling on ice. The suspension will be significantly less viscous and appear more yellow than white when the sonication is complete.

4.3 Protein purification; choosing a method

The protocols for large scale protein induction (*Protocol 7*) and for cell harvesting and extraction (*Protocol 8*) are suitable for both untagged (pET, Section 2.3) and tagged (pGEX and pMAL, Sections 2.5 and 2.6; pET-16b, Section 2.3) recombinant proteins. However, the method of protein purification will vary depending upon the solubility of the recombinant protein as determined in Section 3. It is often valuable to re-check protein solubility in the large scale preparation to ensure that the solubility properties are as determined for the small scale preparation. To do this, centrifuge the induced, lysed culture obtained from *Protocol 8* at 8000 *g* and analyse small samples of the soluble (supernatant) and insoluble (pellet) fractions (dissolved in sample buffer) by SDS–PAGE. Coomassie blue staining of the gel will reveal whether the recombinant protein is located predominantly in the soluble or insoluble fraction.

Soluble proteins are best purified from the crude lysate using tagged protein expression vectors and affinity chromatography (Section 4.3.3 and see Chapter 3), while untagged proteins can be isolated by preparative gel electrophoresis (Section 4.3.2). Insoluble proteins can be easily and consistently purified from the lysate by centrifugation, exploiting the tendency of insoluble recombinant proteins to form intracellular aggregates termed inclusion bodies. However, in order to be useful for immunological purposes, the purified protein in the form of inclusion bodies must then be solubilized as described in the next section.

4.3.1 Inclusion body solubilization and recovery of immunologically active protein

After cell harvesting and lysis, the insoluble recombinant protein, in the form of inclusion bodies, is purified from soluble fractions by centrifugation and washing. The effects of different washing conditions on inclusion body purity are shown in *Figure 7* (lanes 4–9). A mild detergent solution provides for the purest inclusion body preparation (approximately 70–95% pure). When relatively pure, the washed inclusion bodies must then be solubilized for subsequent immunological use. We have found that the use of SDS as a solubilizing agent is suitable for all insoluble antigens we have tested, and is superior to more commonly used agents such as urea and guanidine–HCl. Unlike proteins solubilized in these agents, SDS solubilized proteins generally stay in solution when the SDS is subsequently removed. Furthermore, SDS will not interfere significantly with coupling reactions when the solubilized

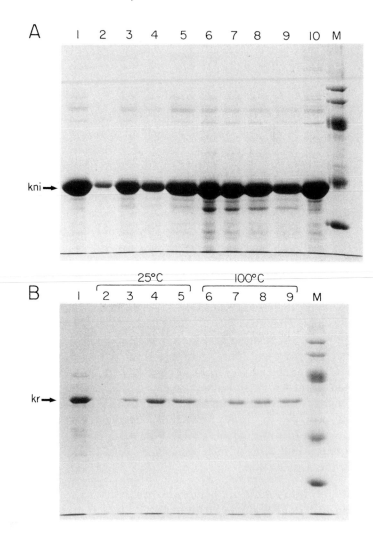

Figure 8. The effects of SDS concentration, protein concentration, and temperature on recombinant protein solubility. (A) The effect of SDS concentration on inclusion body protein solubility was investigated by dissolving *Knirps* protein inclusion bodies in 0.5% (lane 2), 2% (lane 3), 5% (lane 4), or 10% SDS (lanes 1, 5), and heating at 100 °C. The most complete solubilization was with 10% SDS. More dilute protein samples were also prepared and heated at 100 °C for 30 min and once again the 10% SDS samples (lanes 6, 10) were more soluble than the 0.5% (lane 7), 2% (lane 8), or 5% (lane 9) samples. Lane M contained high molecular weight protein size markers (Sigma). (B) The effects of SDS and temperature were examined on the *Krüppel* protein by dissolving inclusion bodies in 0.5% (lanes 2, 6), 2% (lanes 3.7), 5% (lanes 4,8), or 10% SDS (lanes 5, 9), at either 25 °C (lanes 2–5) or 100 °C (lanes 6–9) for 30 min. The higher SDS concentration and temperature were the most effective at dissolving *Krüppel*. Lane M contained high molecular weight protein size markers (Sigma).

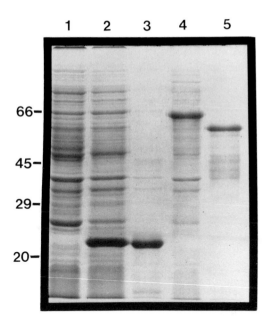

Figure 9. Purification of inclusion body proteins. Three different *Drosophila* proteins were purified from inclusion bodies as described in *Protocol 9*. Lane 1, whole cell lysate of uninduced *E. coli* bearing a pET *achaete* expression construct; lane 2, lysate of induced *E. coli* (*achaete* protein is the prominent band at $M_r \sim$ 22 000); lane 3, *achaete* protein purified from inclusion bodies; lane 4, pET induced *ftz* protein purified from inclusion bodies; lane 5, pET induced *Krüppel* protein purified from inclusion bodies. The sizes and migration positions of protein markers are indicated on the left-hand side of the figure.

protein is used to prepare immunoaffinity columns (Section 5.2.2) We have investigated the influence of SDS concentration, temperature, and protein concentration on the efficiency of solubilization (*Figure 8*) and in general find that high SDS concentrations (greater than 2%) combined with heating provides the most effective dissolution of inclusion bodies. As a standard practice we dissolve the inclusion bodies in two to three volumes of 10% SDS and once dissolved, dilute the sample to 1% SDS, and sequentially dialyse the suspension back to 0.01% SDS (*Protocol 9*). Although we include protease inhibitors in this protocol, these inhibitors are usually unnecessary since insoluble inclusion bodies are relatively inaccessible to *E. coli* proteases.

All insoluble recombinant proteins that we have encountered can be solubilized by this protocol (e.g. see *Figure 9*). However, some inclusion bodies may take considerable patience to solubilize. Alternately heating the protein–SDS sample to 95–100 °C for periods of up to one hour with intermittent pipetting should eventually result in the solubilization of even the most stubborn inclusion bodies. After solubilization and dialysis, the protein–SDS suspension should be quantified by gel electrophoresis and is then suitable for mixing with adjuvant and injection into animals (Section 5.1), as well as for coupling to affinity matrices for purifying antisera (Section 5.2.2). These suspensions can be stored at 4 °C for several months. Residual SDS may come out of solution at this temperature but will dissolve again upon warming to room temperature.

Protocol 9. Solubilization of recombinant proteins from inclusion bodies

Reagents

- RIPA buffer: 0.1% SDS, 1% Triton X-100, 1% sodium deoxycholate in TBS (25 mM Tris–HCl pH 7.5, 150 mM NaCl)
- 10% SDS
- Protease inhibitors (*Protocol 8*)
- PBS
- Sonicated protein suspension (*Protocol 8*)

Method

1. Transfer the lysed, sonicated protein suspension (*Protocol 8*) to a 30 ml Corex centrifuge tube and centrifuge at 8000 g for 10 min at 4 °C to pellet the inclusion bodies.

2. Wash the pellet twice by pipetting or vortexing in fresh, ice-cold RIPA containing 1 mM PMSF, 1–5 μg/ml leupeptin, 1–5 μg/ml pepstatin, and 1–5 μg/ml aprotinin. Re-centrifuge the inclusion bodies after each wash. At this stage the inclusion bodies should be quite pure.

3. After drying, transfer the inclusion bodies to one or more 1.5 ml microcentrifuge tubes such that each tube has no more than 500 μl of inclusion bodies. This is best accomplished using a small metal spatula.

4. Estimate the volume of the inclusion bodies and add 2 vol. of 10% SDS.

5. Solubilize the pellet by gently pipetting up and down with a 1 ml micropipettor. Take care to avoid introducing too many bubbles, which can make solubilization quite messy.

6. Solubilization of the inclusion bodies can be facilitated by heating the sample to 95 °C for periods of up to 1 h followed by more pipetting.

7. Once the inclusion bodies are in solution, dilute the sample with 9 vol. PBS (the SDS concentration is now 1%). The protein should stay in solution.

8. Dialyse the protein solution overnight against a 100-fold volume of PBS containing 0.05% SDS and 1 mM PMSF. Perform this dialysis at room temperature as the SDS will come out of solution at 4 °C.

9. Change the dialysis buffer to PBS containing 0.01% SDS and 1 mM PMSF. Dialyse for several hours to overnight at room temperature.

Some of the protein may precipitate during this dialysis, but most will stay in solution.

10. Store the dialysed solution at 4 °C. Some SDS will come out of solution at this temperature, but it should go back into solution when the sample is warmed up to room temperature.

11. Analyse a small aliquot of the preparation (1 μl) by SDS–PAGE along with protein standards to quantify the final protein concentration and yield as well as the protein integrity.

4.3.2 Preparative SDS–PAGE of untagged proteins

In our experience the majority of recombinant proteins are insoluble when expressed at high levels in E. coli and are then easily purified, by following the inclusion body protocol (*Protocol 9*). In certain instances, however, the recombinant protein is soluble and must be purified in an alternative manner. In these cases we have purified the recombinant protein by preparative SDS–PAGE followed by excision and electroelution of the desired protein band. This procedure (*Protocol 10*) is straightforward and works well when small quantities of protein are needed (< 1 mg), but is rather labour-intensive when large protein quantities are required (> 10 mg). Further, the purity of the recombinant protein can be low ($\approx$ 50%), if induction is poor and E. coli proteins co-migrate with the recombinant protein. Because of these problems we recommend, if possible, that inserts coding for soluble proteins be cloned into tagged protein expression vectors and that the recombinant protein be purified via the suitable affinity column (see Sections 3, 4.3.3, and Chapter 3). Such protein expression and purification systems allow large quantities of quite pure protein to be obtained in a more efficient manner.

Protocol 10. Purification of soluble recombinant proteins by preparative SDS–PAGE

Equipment and reagents

- Sonicated protein suspension (*Protocol 8*)
- 10% polyethylenimine (Sigma)
- Centricon-30 microconcentrator (Amicon)
- 2 × sample buffer (*Protocol 3*)
- 1 × SDS running buffer: 1% SDS, 50 mM Tris base, 385 mM glycine
- 16 cm × 18 cm glass gel plates
- 1.5 mm (thick) × 16 cm (tall) × 1 cm (wide) gel spacers

- 7.5 or 12% SDS–polyacrylamide gel and vertical SDS–PAGE apparatus
- Coomassie blue stain: 0.6% Coomassie blue R-250, 20% methanol, 20 mM Tris–HCl pH 7.5
- 20 mM Tris–HCl pH 7.5
- Horizontal gel electrophoresis apparatus (optional—for electroelution of purified protein)

Method

1. Transfer the lysed, sonicated protein suspension to a 30 ml centrifuge tube and centrifuge at 8000 *g* for 10 min at 4 °C to remove cellular debris. Keep the supernatant.

Protocol 10. *Continued*

2. (Optional) Removal of chromosomal DNA by polyethylenimine precipitation. Slowly stir in polyethylenimine to a final concentration of 0.5%; stir for 20 min at 4 °C. Centrifuge for 15 min at 7500 *g*. Keep the supernatant.

3. Concentrate the supernatant two- to threefold using a Centricon-30 microconcentrator following the manufacturer's guide-lines.

4. Add 500 µl of concentrated supernatant to 500 µl of 2 × sample buffer and heat to 95 °C for 5 min.

5. Fractionate 1 ml of sample using a 1.5 mm (thick) × 16 cm (tall) × 18 cm (wide) preparative 7.5% SDS–PAGE gel. (If the recombinant protein is < 30 kd it may be better to use a 12% PAGE gel.) Electrophorese until the bromophenol blue dye has migrated to the end of the gel.

6. Stain the gel in Coomassie blue stain for 5 min. Destain for 30–60 min using several changes of 20 mM Tris–HCl pH 7.5.

7. Excise the desired band. At this point, the gel slice can be pulverized and then emulsified with Freund's adjuvant. The resulting mixture can be used as an immunogen without further treatment. However, we prefer to electroelute the recombinant protein and then use it as an immunogen or for other purposes, e.g. ELISA analysis. To achieve this, continue with steps 8–11.

8. Cut the gel slice into small pieces (2 mm × 2 mm). Place these pieces into a dialysis bag with 5–10 ml of 1 × SDS running buffer. Place the bag into a horizontal gel electrophoresis apparatus and submerge it with 1 × SDS running buffer.

9. Electroelute the recombinant protein for 2–3 h at ≈ 90 mA.

10. Collect the eluate. Place in new dialysis bag and dialyse against three changes of PBS (> 1 litre each time).

11. Small aliquots (10 µl) can be analysed for purity and quantity by SDS–PAGE as in *Protocol 3*.

4.3.3 Affinity chromatography of tagged proteins

In cases where an induced protein is both soluble and affinity tagged, the simplest purification method is to isolate soluble protein by affinity chromatography. Large scale lysates of soluble *E. coli* proteins are produced and affinity purified by metal chelate affinity chromatography on nickel columns in the case of poly-histidine tagged proteins (see Chapter 3, Section 5.1 for relevant protocols), on glutathione–Sepharose columns for pGEX expressed proteins (see Chapter 3, Section 5.2, or ref. 16 for relevant protocols), or on

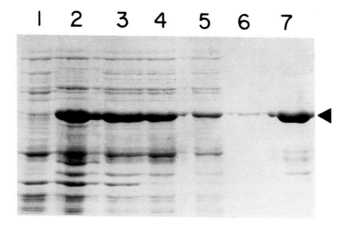

Figure 10. Affinity purification of a pMAL fusion protein on amylose resin. The *Precis coenia* decapentaplegic protein is shown at different stages of purification as a pMAL-C fusion protein. Protein samples from uninduced (lane 1) and induced (lane 2) one litre cultures (*Protocol 7*) show accumulation of fusion protein after induction (*arrow*). The culture was pelleted and sonicated (*Protocol 8*) and clarified by centrifugation (10 000 *g*, 30 min). Extracts of total protein (lane 3) and soluble protein (the clarified lysate, lane 4) demonstrated that most of the fusion is soluble. The soluble fraction was applied to an amylose column (methods provided in ref. 16). The flow-through fraction is depleted of most of the fusion (lane 5), which is eluted in highly pure form in sonication buffer containing 10 mM maltose (lanes 6 and 7).

amylose columns for pMAL fusions (see ref. 16 for detailed protocols). *Figure 10* shows an example in which a soluble pMAL fusion protein is purified to near homogeneity on an amylose column.

5. Polyclonal antibody production and purification

Using the methods described in Section 4, recombinant proteins may be obtained in a variety of forms and in differing amounts and degrees of purity. The best situation is where the amounts of antigen are not limiting because, in our experience, satisfactory antibodies can be made to denatured proteins containing impurities so long as sufficient (milligram) quantities of proteins are available for raising high titres of antibodies and for purifying the resulting polyclonal antiserum.

Given satisfactory quantities of antigen, the next issue is the choice of host to be immunized. Unless there is prior experience with the recombinant protein and very large quantities of antibodies are needed, we would discourage the use of larger mammals (e.g. goats, sheep) or laying hens for antibody production. For most research applications, the choices usually boil down to rabbits, rats, and mice. Clearly, for monoclonal antibody production, the latter is the obvious choice. However, we have not found monoclonals to be necessary as first generation reagents. Normally, polyclonal antibodies made in rabbits or rats will suffice for most applications and can be produced and handled more easily by the less experienced immunologist. The main differences between the use of rabbits and rats in polyclonal antibody production are a function of their different body sizes. For a given quantity of antigen, it is easier to attain a higher immune response in rats because of their lower body weight, that is, one milligram of a protein is approximately a 4 mg/kg dose in an adult rat but only a 0.5 mg/kg dose in a rabbit. Therefore, the titre of antibody per millitre of antiserum is likely to be higher in a rat and may be high enough that the antiserum can be used without any purification. On the other hand, one can routinely obtain much more antiserum from a rabbit, and even if it is of a lower titre, affinity purification will result in the maximum level of antibody potency. Typically, one will choose to immunize a group of rats (perhaps four to six) with a given antigen and/or a pair of rabbits. The individual rat sera can be checked periodically for desired characteristics and the unsatisfactory sera discarded. We prefer, however, to immunize rabbits with the largest quantities of antigen available and to purify the much larger volumes of sera by affinity chromatography to create long-lasting stocks of fully-characterized, purified, antigen-specific antibodies. In the systems described here, it is common to isolate tens of milligrams of recombinant protein which can be used to produce several milligrams of specific antibody within six to eight weeks. In the course of a four to six month immunization schedule, one pair of rabbits will yield approximately 300 ml of antiserum containing 50–150 mg of specific antibody. For most immunoassays (Western blotting, immunohistochemistry, immunoprecipitation) this represents sufficient antibody for tens of thousands of assays. (The only downside of this approach is that colleagues in the field tend to rely upon these supplies once they are available.)

5.1 Immunization schedule

Rabbits should be immunized with the maximum amount of antigen that can be readily obtained, up to about 10 mg per injection. For proteins that are expressed at the level of $\geq$ 30 mg/litre of culture and which form inclusion bodies that can be processed as in Section 4.3.1, this should pose little problem. For proteins that are poorly expressed (e.g. only a few mg/litre of culture) or are found in the soluble fraction of the cell lysate and possess no affinity tag, this is more difficult. For a 5–6 lb rabbit one should try to immunize with no less than 0.5 mg of protein per injection.

To prepare the injection, mix 0.5–10 mg of recombinant protein in $\leq$ 1 ml of saline or PBS (the SDS contained in inclusion body preparations is not a problem) with 1.25 volumes of complete Freund's adjuvant (Gibco/Life Technologies) using two Luer/lock tip syringes with a stainless steel connector (this connector can be made by brazing two 18 gauge needles together near their base.) Emulsify the antigen completely (it may require several minutes and dozens of strokes) so that the suspension is of a firm consistency and stable and inject at several subcutaneous sites. Repeat the injections with 0.5–10 mg of antigen in incomplete adjuvant 14 and 21 days later and take the first bleeding ($\leq$ 20 ml from the ear vein) on day 28. Boost the rabbits every 21–28 days with 0.5–10 mg of antigen and bleed seven to ten days after each boost. Although animal dependent, expect the specificity of the antiserum to increase with time, reaching maximal levels by anywhere from two months to up to a year after initial injection. Once the antiserum appears satisfactory, blood may be collected weekly as long as the animal is healthy. When satisfactory quantities of antisera of sufficient specificity have been obtained and while the titre is still high (boosting, if necessary), exsanguinate the rabbit under anaesthesia by heart puncture.

5.2 Antiserum purification

For many applications, it is necessary or desirable to utilize affinity purified antibodies specific to a recombinant protein. Affinity purification of pooled sera allows for the preparation of larger batches of antibody reagents with consistent properties and overcomes the lower and variable titres of whole antisera. The procedures below detail methods for fractionating the whole antiserum against a recombinant protein to obtain antibodies specific for the protein of interest.

5.2.1 Antiserum preparation from whole rabbit blood

Collect the blood into a 50 ml plastic disposable centrifuge tube. Apply a wooden applicator stick around the wall of the tube to free any clots that have formed and place the stick into the centre of the clot forming in the centre of the tube. Place the blood in a refrigerator (4 °C) overnight. The next day, remove the clot attached to the applicator stick and centrifuge the remaining liquid at 13 000 g for 10 min. Decant or pipette the clarified antiserum into sterile tubes and add 20% sodium azide (w/v) (**NB**: this reagent is highly toxic!) to a final concentration of 0.02%. Store at 4 °C if to be used within a few weeks, or if not to be processed in that time, freeze until needed at −20 °C.

5.2.2 Preparation of affinity columns

One of the principal reasons for developing the procedures described in Section 4 for solubilizing proteins from inclusion bodies was to facilitate their covalent attachment to affinity matrices used in antibody purification. While many denaturants can solubilize inclusion body proteins, some other more common solvents, such as urea and guanidine, contain free amino groups that will interfere with typical covalent coupling chemistries. In addition, the removal of these agents is often accompanied by precipitation of a significant portion of the solubilized protein. What was needed was a general method for solubilizing virtually any protein that would not interfere with coupling chemistries; the SDS solubilization technique (*Protocol 9*) has worked for all proteins we have examined.

Advances in affinity matrix design have led to the development of some simple, ready-to-use resins that provide high coupling efficiencies and stable attachment chemistries. In particular, aldehyde-activated resins that react with free amino groups appear to have superior characteristics. In order to couple solubilized recombinant proteins, it is desirable to remove most of the SDS by dialysis from the protein solution since high concentrations of SDS (0.5% w/v) will reduce coupling efficiency. A column capable of purifying several milligrams of antibody can be prepared with just a few milligrams of recombinant protein (*Protocol 11*). Typical coupling efficiencies, which must be monitored by gel electrophoresis (the SDS and nucleic acid in the inclusion body sample interferes with most other assays), typically range from 40–90%. These columns can be utilized for dozens of cycles and are stable for at least one year.

Protocol 11. Preparing an affinity column

Equipment and reagents

- 5 ml protein solution containing 1–3 mg/ml recombinant protein (*Protocol 9*)
- 5 ml Actigel-A (Sterogene Biochemicals)
- PBS
- 1 M NaCNBH$_3$
- Scintered glass funnel
- Small affinity column (e.g. Econo-column, Bio-Rad)

Method

1. Wash 5 ml of the Actigel-A resin on a scintered glass funnel (do not allow the resin to dry!) with 50 ml of PBS and filter it to form a moist cake.

2. Add the washed Actigel-A resin to 5 ml of the protein solution (retaining 0.1 ml for later analysis) in a 15 ml polypropylene centrifuge tube.

3. Add 1 ml of 1 M NaCNBH$_3$, and incubate with gentle rotation on an orbital shaker for at least 4 h at room temperature or overnight at 4 °C.

4. Filter the resin on a scintered glass funnel (do not allow the resin to dry!) and retain the filtrate for further analysis. Wash the resin with 50 ml of PBS and transfer the resin to the small affinity column. Alternatively, one may let the resin settle before washing and assay a sample of the supernatant to assess coupling efficiency.

5. Analyse 1, 5, and 10 μl samples of both the initial protein solution and the post-coupling filtrate by SDS–PAGE, and compare the relative Coomassie blue staining intensity by inspection or by densitometry.

5.2.3 Purification of antibodies specific for recombinant proteins

Rabbit antisera raised against milligram doses of recombinant antigens typically reach titres of 100–500 µg of specific antibody per millilitre of serum within 30–60 days of the beginning of the immunization regimen. Thus, from a typical 20 ml batch of serum from a pair of rabbits, one can expect to obtain 2–10 mg of specific antibodies by affinity purification (*Protocol 12*). Columns prepared as described in *Protocol 11* will generally have this capacity since we have found, on average, that 1 mg of coupled protein will bind 0.5–1.0 mg of antibody. Affinity purification is directly scalable so larger columns will have proportionally greater capacities and yields.

Protocol 12. Affinity purification of antibodies

Equipment and reagents

- UV absorbance monitor/recorder
- Protein affinity column (*Protocol 11*)
- 10–50 ml of antiserum (clarified by centrifugation if necessary)
- PBS
- BBS-Tween buffer (*Protocol 6*)
- 4 M guanidine–HCl, 10 mM Tris pH 8.0
- 20% (w/v) NaN$_3$
- Dialysis tubing ($\leq$ 10 000 M_r cut-off)

Method

1. Connect the column to the flow cell of a UV absorbance monitor equipped with 280 nm filters and establish the baseline with PBS flowing through the column.

2. Strip non-covalently bound protein from the column by applying two column volumes of 4 M guanidine–HCl, 10 mM Tris pH 8.0 to the column. Re-equilibrate the column with several volumes of PBS.

3. Apply the antiserum to the column at a flow rate of approximately 1 ml/min. Collect and save the effluent.

4. Wash the column with 1 vol. of PBS, then with BBS-Tween until the absorbance returns to baseline. Then wash with at least two column volumes of PBS.

5. Elute the antibody with 4 M guanidine–HCl, 10 mM Tris pH 8.0 and collect the entire peak fraction on ice.

6. Re-equilibrate the column with 5 vol. of PBS, adding NaN$_3$ to 0.02% in the last volume.

7. Immediately dialyse the antibody peak fraction against three changes of 1 litre of PBS at 4 °C over 12 h. Centrifuge at 10 000 g for 10 min to remove any aggregates.

8. Quantitate the antibody peak fraction by measuring OD$_{280}$; a 1 mg/ml solution of IgG has A$_{280}$ of 1.4.

Since the recombinant protein utilized in the immunization and coupling procedures usually contains some *E. coli* proteins, it may be desirable to remove any *E. coli* reactive antibodies by passage of the affinity purified antibody through a matrix containing proteins solubilized from an *E. coli* strain expressing a heterologous recombinant protein or from a mock inclusion body preparation (*Protocol 13*). In the case of fusion proteins, antibodies specific to the carrier protein are removed in the same manner, using a column containing a total protein lysate made from an induced culture containing the parent vector. Such columns are easily made from an induced 500 ml culture (*Protocol 7*). The culture is lysed in inclusion body sonication buffer (*Protocol 8*), clarified by centrifugation (8000 *g*, 10 min) and directly coupled to 10 ml of Actigel resin (*Protocol 11*). This will provide an absorption column containing up to 100 mg of *E. coli* protein, since the capacity of the resin is high (greater than 10 mg/ml). Cross-absorption of the affinity purified antibody readily removes the *E. coli* reactive protein as assayed by Western blotting (*Figure 11*). It is useful to keep the antibody pools that are reactive to the maltose binding protein. An affinity column containing these antibodies can be used to immunoaffinity purify (5) pMAL fusions that are unable to bind amylose resin (approximately 30% of all fusions).

Protocol 13. Cross-absorption of purified antibodies

Equipment and reagents

- UV absorbance monitor/recorder
- Affinity purified antibody (*Protocol 12*)
- PBS
- 5 ml affinity column with either heterologous inclusion body protein (*Protocol 11*) or total *E. coli* protein

- 20% (w/v) NaN₃
- 4 M guanidine–HCl, 10 mM Tris pH 8.0 (for column regeneration)

Method

1. Establish the baseline for the absorbance monitor at high sensitivity with PBS flowing through the column.

2. Apply 1–10 mg of affinity purified antibody equilibrated in PBS to the affinity column. Wash with PBS until the entire protein-containing effluent is collected. Quantify the eluted antibody by OD_{280}.

3. Add 20% NaN₃ (w/v) to the effluent to 0.02% for long-term storage at 4 °C.

4. Regenerate the column by stripping it with 4 M guanidine–HCl, 10 mM Tris pH 8.0, followed by washing with five column volumes of PBS.

5.3 Special techniques to optimize antibody reactivity

Antibodies prepared as in Section 5.2 may be employed in a wide variety of qualitative or quantitative immunoassays to detect recombinant or native

A   **B**

1   2   1   2

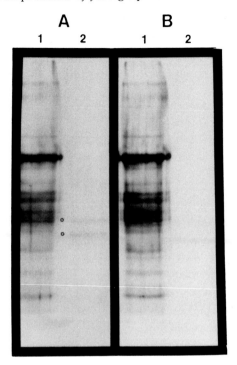

Figure 11. Cross-absorption of affinity purified antibodies to remove *E. coli* reactive impurities. Immunoreactivity of affinity purified rabbit anti-*Krüppel* antibodies (A) before and (B) after cross-absorption on a column containing a heterologous inclusion body preparation. Lane 1, *Krüppel* protein; lane 2, a heterologous inclusion body preparation. Note the two protein bands in lane 2, panel A (marked with *open circles*) that are not reactive with the cross-absorbed antibody (lane 2, panel B).

proteins. Since the inclusion body solubilization scheme denatures proteins and no effort is made to re-fold polypeptides into their native state, it is likely that the bulk of the antibodies are raised against sequential and not conformational antigenic determinants. This may compromise the reactivity of the antibodies with native proteins which is especially important in immunohistochemical applications employing fixed intact cells, tissues, or whole organisms. One way to increase the reactivity of the antibody is to use more gentle means of disrupting antibody–antigen complexes formed during affinity chromatography. Typically, the use of low pH or hydrogen bond disrupting agents will reduce the specific activity of an antibody population by more than 50%. We have found that gentler agents can increase the specific activity of the purified antibody (*Protocol 14*) and provide qualitatively superior reagents for sensitive applications such as immunohistochemistry. While gentle elution is often not necessary with hyperimmune antisera and does not remove all of the bound antibody from the affinity matrix, it is worth exploring when antibodies appear to have lower signal-to-noise ratios than may be desired.

Protocol 14. Gentle elution of antibodies from affinity columns

Equipment and reagents

- TBS: 25 mM Tris–HCl pH 7.5, 150 mM NaCl
- Actisep elution medium (Sterogene Bio-chemicals)
- All other reagents are described in *Protocol 12*

Method

1. Establish the baseline for the absorbance monitor as in *Protocol 12*.
2. Apply antiserum, PBS, and BBS-Tween as in *Protocol 12*.
3. Wash the column with at least 2 vol. of TBS. [a]
4. Apply Actisep elution medium until the eluate peak reaches a maximum on the absorbance monitor. To ensure maximal elution, stop the column for 90 min.
5. Resume elution and collect the eluate until a *new* baseline is reached (Actisep absorbs at 280 nm).
6. Wash the column with TBS until the original baseline is reached. Then strip remaining antibody from the column with 4 M guanidine–HCl, 10 mM Tris pH 8.0. Finally, re-equilibrate the column with PBS.
7. Dialyse the eluate (step 5) against several changes of TBS or PBS at 4°C for 12–16 h. Be sure to allow room in the dialysis bag for the sample to increase several-fold in volume!
8. Quantify the antibody yield by measuring OD_{280}.

[a] PBS in *Protocol 12* is replaced with TBS here since Actisep is not compatible with PBS.

5.3.1 Purification of cross-reactive antibody subpopulations with homologous proteins from other species

In many cases it is desirable to use an antibody to examine protein expression from a number of different species. Indeed, polyclonal antibody pools raised against a protein from one species often cross-react with the target protein in other species. Unfortunately, the strength of this cross-reactivity usually decreases rapidly as species divergence increases, due to protein divergences at the amino acid or conformational level that remove cross-reactive epitopes. We describe here a technique to prepare interspecifically reactive antibody preparations, by selectively purifying antibodies reactive to evolutionarily conserved epitopes.

Polyclonal antibodies are comprised of pools of antibodies reactive against various epitopes present in the target protein. These epitopes may be linear (specific to a sequence of about six amino acids in the protein) or conformational. In the case of polyclonal antibodies raised against solubilized inclusion body protein, most of the purified antibodies will recognize linear epitopes due to the fact that the solubilized protein will not be properly folded. In an evolutionarily conserved protein, a percentage of these linear epitopes may be well conserved between different species, essentially representing stretches of absolutely conserved amino acids. Of course, antibody reactive epitopes are distributed non-randomly within proteins, such that it is difficult to predict what fraction, if any, of a polyclonal antibody pool will be interspecifically reactive, even if sequence analysis has demonstrated the degree of protein conservation.

If a cDNA encoding the protein of interest is available from a second species, cross-reactive antibodies may be purified from existing antibody pools. The heterologous protein is expressed and purified as described in Sections 3 and 4, and a protein affinity column prepared (*Protocol 11*). Antiserum, reactive against the first species, is affinity purified against this column (*Protocol 12*), and *E. coli* reactive antibodies are removed (*Protocol 13*). The resultant antibody preparation will contain only those antibodies that are reactive against evolutionarily conserved epitopes. This enrichment may dramatically enhance antibody reactivity to other species (J. A. Williams and S. B. Carroll, unpublished observations).

Acknowledgements

We thank Allen Laughon, Tadashi Uemura, Dusan Stanojovic, Mike Levine, Gary Struhl, and Claude Desplan for some of the constructs utilized in the protein expression experiments; Dave Keys for critical review of the manuscript; Leanne Olds for help with the illustrations; and Jamie Wilson for assistance with preparation of the chapter. We also acknowledge the rabbits, mice, and chickens for their parts in these experiments and for the valuable reagents they produced for our research. Jim Williams was supported by a Medical Research Council (Canada) Postdoctoral Fellowship and James Langeland is a Howard Hughes Medical Institute Predoctoral Fellow. This work was supported by National Science Foundation grant DCB-8857124, by the Shaw Scientists Program of the Milwaukee Foundation, and the Howard Hughes Medical Institute.

References

1. Marston, F. A. O. (1987). *In DNA cloning: a practical approach* (ed. D. Glover), Vol. III, pp. 59–88. IRL Press, Oxford.
2. Marston, F. A. O. (1986). *Biochem J.*, **240**, 1.

3. Riggs, P., Hoey, T., Smith, D. B., and Corcoran, L. M. (1990). In *Current protocols in molecular biology* (ed. F. M. Ausubel, R. Brent, R. E. Kingston, D. D. Moore, J. G. Seidman, J. A. Smith, and K. Struhl), Chap. 16.6, supplement 10. Green Publishing Associates and Wiley-Interscience, NY.
4. Rüther, U. and Müller-Hill, B. (1983). *EMBO J.*, **2**, 1791.
5. Carroll, S. B. and Laughon, A. S. (1987). *In DNA cloning: a practical approach* (ed. D. Glover), Vol. III, pp. 89–111. IRL Press, Oxford.
6. Koerner, T. J., Hill, J. E., Myers, A. M., and Tzagoloff, A. (1991). In *Methods in enzymology* (ed. C. Guthrie and G. R. Fink), Vol. 194, pp. 477–90. Academic Press, San Diego.
7. Löwenadler, B., Nilsson, B., Abrahmsen, L., Moks, T., Ljungquist, L., Holmgren, E., Paleus, S., Josephson, S., Phillipson, L., and Uhlen, M. (1986). *EMBO J.*, **5**, 2393.
8. Smith, D. B. and Johnston, K. S. (1988). *Gene*, **67**, 31.
9. Studier, F. W., Rosenberg, A. H., and Dunn, J. J. (1990). In *Methods in Enzymology* (ed. D. V. Goeddel), Vol. 185, pp. 60–89. Academic Press, San Diego.
10. Krueger, J. K., Kulke, M. H., Schutt, C., and Stock, J. (1989). *BioPharm.*, **2**, 40.
11. Studier, F. W. and Moffat, B. A. (1986). *J. Mol. Biol.*, **189**, 113.
12. Guan, C., Li, P., Riggs, P. D., and Inouye, H. (1987). *Gene*, **67**, 21.
13. Maina, C. V., Riggs, P. D., Grandea, A. G. III, Slatko, B. E., Moran, L. S., Tagliamonte, J. A., McReynolds, L. A., and Guan, C. (1988). *Gene*, **74**, 365.
14. Johnston, G. I., Cook, R. G., and McEver, R. P. (1989). *Cell*, **56**, 1033.
15. Sambrook, J., Fritsch, E. F., and Maniatis, T. (ed.) (1989). *Molecular cloning, a laboratory manual*. Cold Spring Harbor Press, Cold Spring Harbor, NY.
16. Ausubel, F. M., Brent, R., Kingston, R. E., Moore, D. D., Seidman, J. G., Smith, J. A., and Struhl, K. (ed.) (1989). *Current protocols in molecular biology*. Greene Publishing Associates and Wiley-Interscience, NY.

3

Purification of over-produced proteins from *E. coli* cells

REINHARD GRISSHAMMER and KIYOSHI NAGAI

1. Introduction

Many biologically interesting proteins are present naturally in small quantities and are not amenable for structural and biochemical characterizations. The molecular mechanisms of gene expression in *Escherichia coli* became well understood in the late 1970s, allowing the over-production of foreign proteins from their cloned genes. A large proportion of new protein structures reported in 1990 and 1991 were derived from heterologously over-produced proteins (1) and this unequivocally demonstrates that the technique of over-production has now become an essential tool for the study of protein structure and function. We mainly focus in this chapter on the purification of over-produced proteins from *E. coli* cells. It is a well-known fact that over-produced proteins are not always soluble and active in *E. coli* and often aggregate as insoluble inclusion bodies. As re-folding of proteins is difficult, it is desirable to produce proteins in a soluble and active form. We first describe how inclusion body formation can sometimes be avoided. We then detail purification methods for both soluble and insoluble proteins. Each protein has its own 'personality' and the methods described in this chapter should be regarded as a collection of experiences which may direct readers to the successful purification of their own proteins.

2. Factors that affect physical properties of over-produced proteins in *E. coli* cells

Mechanisms of inclusion body formation are not fully understood but are probably due to the aggregation of incompletely folded polypeptides. Inclusion bodies often contain cellular components such as ribosomes, RNA polymerase, and DNA that co-precipitate with the nascent polypeptide chains (2). The physical properties of inclusion bodies are therefore variable, depending on the chemical nature of the over-produced protein and its interaction with cellular components. Some inclusion bodies are readily solubilized in urea; these proteins can be purified by ion-exchange chromatography in the presence of urea. Others interact strongly with cellular components and cannot easily be purified. Whether a foreign protein forms inclusion bodies depends not only on the chemical composition of the polypeptide but also on various factors such as culture media, growth temperature, expression vectors, and/or the gene sequence. Inclusion body formation can sometimes be avoided by changing these variables.

2.1 Codon usage

Genetic codes for some amino acids are highly degenerate (e.g. six different codons are used for serine and arginine) and these codons are not equally used in each organism. Ikemura (3) discovered that highly expressed *E. coli* genes preferentially use codons recognized by abundant tRNA species. He proposed that the availability of charged tRNAs may be a rate-limiting factor of protein synthesis and that the production of foreign proteins can be boosted by the use of codons for abundant tRNA species. The limited availability of charged tRNAs has been suggested to cause pausing of polypeptide elongation which in turn results in misfolding and rapid degradation of the nascent polypeptide chain as well as misincorporation of amino acids (4). We and many others have observed that the use of chemically synthesized genes with codons optimized for high level expression in *E. coli* can result in the production of soluble proteins (5, 6) when the use of natural cDNAs failed in protein production or resulted in the formation of inclusion bodies. Because chemical synthesis of oligonucleotides has become reliable and relatively inexpensive, it is worthwhile to synthesize genes using codons optimal for expression in *E. coli*. It must be noted, however, that the use of chemically synthesized genes does not always lead to the production of soluble proteins.

2.2 Expression vector

The use of strong promoters often results in the formation of inclusion bodies. As large amounts of protein accumulate rapidly in the cell, incompletely folded polypeptides aggregate more readily. The use of weaker promoters sometimes favours the production of proteins in soluble form. Some proteins become insoluble simply by increasing the copy number of the expression vector (N. Komiyama, unpublished data).

2.3 Growth temperature

Some proteins become soluble by changing the growth temperature from 37 °C to 30 °C or below (7). This is not only due to the increased physical stability of many proteins at lower temperature but also because the physiological state of *E. coli* cells and the rate of protein synthesis and protein folding are different at lower temperature.

2.4 Growth media and aeration

Both culture media and aeration affect the growth rate and the metabolic state of the cells. We have found that carp haemoglobin becomes soluble when cell growth is carried out under poor aeration (N. Komiyama, unpublished data), whilst a protein component of the small nuclear ribonucleoprotein complex becomes soluble only under good aeration (J. Avis, unpublished data).

2.5 Oligomer formation

Human haemoglobin consists of two pairs of α- and β-subunits. When the α-globin gene alone is cloned into an expression vector, no protein production is detected, probably due to rapid degradation. β-Globin is found in inclusion bodies when it is expressed on its own. When both genes are arranged in a polycistronic operon, active and soluble haemoglobin is produced (6). This example clearly demonstrates that subunits of oligomeric proteins can be stabilized by co-expressing multiple components. Other proteins known to interact strongly have also been produced functionally only when they were co-expressed.

3. Purification of soluble recombinant proteins

The mode of gene expression affects the localization of the target protein produced. Recombinant proteins may be located either in the cytoplasm of *E. coli* or translocated across the cytoplasmic membrane into the periplasmic space. Eukaryotic genes may be expressed either direct or as in-frame fusions with synthetic sequences (for example affinity tags) or bacterial genes. The target protein may be soluble or aggregated in an insoluble form (inclusion bodies) (8).

This section will focus on general aspects of the purification of soluble proteins, whereas a later section specifically deals with the purification of soluble fusion proteins by affinity chromatography. Protein purification from inclusion bodies is described in Section 4.

3.1 Preparation of total cell lysate

The preparation of lysates represents a critical step in the purification process. This step influences the total quantity of the desired protein recovered, its biological activity, its integrity by avoidance of proteolytic degradation, and its association with other cellular components. Several factors determine the success of a lysis method. The degree of lysis and functionality of the recombinant protein may be influenced by strain differences and the growth phase at which the cells are harvested, whether the cells

61

are processed immediately or are frozen, the presence of protease inhibitors, and the choice of buffers. A trial and error approach is often required to optimize lysis conditions.

Several methods have been used for the preparation of total cell extracts. These include enzymatic lysis of bacterial cells in the presence of lysozyme/EDTA and mechanical disruption techniques (bead mill, French press, sonication). *Protocol 1* describes the disruption of *E. coli* cells by sonication.

Protocol 1. Disruption of *E. coli* cells by sonication

Equipment and reagents

- Wash buffer and lysis buffer: different buffers are used depending on the respective purification protocol. The lysis buffer may contain optionally 10 mM EDTA,[a] 10 mM 2-mercaptoethanol,[a] and protease inhibitors. The buffer compositions will be those that are listed in the respective purification protocols.

- Sonicator with a 5 mm diameter probe (Branson, Heat Systems)
- Sorvall GS3 rotor or equivalent
- Beckman 70 Ti rotor or equivalent

Method

1. Centrifuge 1 litre of an *E. coli* cell suspension for 10 min at 4 °C in a Sorvall GS3 rotor at 8000 g to collect the cells. Discard the supernatant.

2. Resuspend the cells in 250 ml ice-cold wash buffer (suitable for the subsequent purification step) and centrifuge as in step 1.

3. Resuspend the cell pellet in 25 ml ice-cold lysis buffer (suitable for the subsequent purification step) and transfer the suspension into a glass beaker.

4. Place the beaker in an ice water-bath. Sonicate with short bursts (10 sec) to avoid heating the extract. Allow the cell extract to cool between the single bursts (if the sonicator has a 'pulse' mode, use a 50% duty cycle). Lysis is complete when the cloudy cell suspension becomes translucent.

5. Centrifuge the extract at 35 000 r.p.m. (100 000 g) at 4 °C using a Beckman 70 Ti rotor.

6. Carefully transfer the clear supernatant to a beaker, keep cold, and proceed with purification.

[a] Note that chelating or reducing agents are not always compatible with the purification procedures described in Section 5. For details, see the individual purification protocols.

Several aspects must be considered to obtain optimal results. During sonication, mechanical energy is converted to heat and therefore precautions must be taken not to warm up the cell suspension. Foaming must be avoided to prevent surface denaturation and oxidation of the recombinant protein. Excessive sonication may produce fine cellular debris which may hinder subsequent processing. Therefore a balance has to be found between a minimum time of sonication and maximal cell lysis. To estimate the efficiency of cell lysis, intact cells are counted under the light microscope before and after the lysis procedure.

To aid the disruption of cells by sonication, the sample may be subjected to cycles of freezing in dry ice followed by thawing at 20 °C before sonication. Freezing weakens the cells so that lysis occurs more readily. However, recombinant proteins are more likely to suffer degradation and denaturation with this method. Alternatively, the cell suspension may be incubated with 10 mM EDTA (see footnote in *Protocol 1*) and 0.1 mg lysozyme/ml on ice for 30 minutes followed by sonication.

For many unstable proteins, most of the proteolytic degradation happens during cell breakage. Therefore, it is best to proceed quickly and to always keep the cells and the extract chilled. Addition of EDTA (10 mM final concentration) (see footnote in *Protocol 1*) in the lysis buffer inhibits metal ion-dependent proteases. Other protease inhibitors such as phenylmethylsulfonyl fluoride (PMSF), iodoacetamide (reacts with cysteine residues, and therefore should be avoided if free cysteine residues of the target protein are to be kept unmodified), benzamidine, bacitracin, and/or leupeptin may be included on a case-to-case basis. Disulfide bonds usually do not form intracellularly in *E. coli* due to the reducing environment of the cytoplasm. Therefore, 2-mercaptoethanol (10 mM final concentration) (see footnote in *Protocol 1*) may be included in the lysis buffer to prevent interchain disulfide bond formation upon lysis. Care must be taken that the composition of the lysis buffer does not interfere with the biological activity of the target protein or further purification procedures.

3.2 Preparation of periplasmic protein fraction

Depending on the target protein of interest, proper disulfide bridge formation may be required for its full physical stability and biological activity. The cytoplasm of *E. coli* is a reducing environment. Therefore the target protein has to be directed across the *E. coli* inner membrane into the periplasmic space. *E. coli* periplasmic proteins are normally only 4% of the total cell protein and so less extensive purification of the recombinant protein is required than for proteins located in the cytoplasm. However, expression levels achieved in the cytoplasm usually exceed the levels obtained with target proteins directed into the periplasm.

The release of the contents of the periplasm is described in *Protocol 2*, based on the osmotic shock procedure.

Protocol 2. Preparation of the periplasmic protein fraction of
E. coli by cold osmotic shock

Equipment and reagents

- Sucrose buffer[a]: 50 mM Tris–HCl pH 7.4,
  1 mM EDTA, 20% sucrose (w/v)
- 5 mM MgCl$_2$
- Sorvall GS3 rotor or equivalent

Method

1. Centrifuge 1 litre of an *E. coli* cell suspension for 10 min at 4°C in a Sorvall GS3 rotor at 8000 *g* to collect the cells. Discard the supernatant.

2. Resuspend the cells in 250 ml ice-cold sucrose buffer. Incubate for 10 min on ice with shaking or stirring.

3. Centrifuge as in step 1. Remove the supernatant.

4. Resuspend the pellet in 100 ml ice-cold 5 mM MgCl$_2$.[b] Shake or stir for 10 min in an ice bath.

5. Centrifuge at 8000 *g* for 15 min in a GS3 rotor at 4°C. Save the supernatant, which is the cold osmotic shock fluid. If the supernatant is turbid, re-centrifuge and filter through a 0.2 μm filter.

[a] Note that chelating agents are not always compatible with a respective purification procedure and may be left out. For details, see the individual purification protocols.
[b] Mg^{2+} stabilizes the sphaeroplasts and therefore prevents contamination of the periplasmic fraction with cytoplasmic proteins.

3.3 Chromatographic purification of soluble proteins

A wide variety of purification procedures for soluble proteins has been developed in the past. These include methods such as precipitation, as well as chromatographic and electrophoretic procedures. Specialized techniques have been employed for purification of multienzyme complexes, glycoproteins, or DNA binding proteins. It is beyond the scope of this chapter to describe the various methods in detail. Instead, general aspects of the purification of soluble proteins by chromatographic methods will be considered. A detailed guide to protein purification has been published recently (9).

3.3.1 Ion-exchange chromatography

Proteins contain both positive and negative charges resulting from the ionization of acidic and basic amino acid residues. The pK$_a$ value of the ε-amino group of a free lysine is around 10 and that of the guanidinium group of a free arginine is about 12.5. The side-chains of these amino acids are fully positively-charged below pH 9. The pK$_a$s of free glutamic acid and aspartic acid side-chains and the C-terminal carboxyl groups are around 4. These groups are fully negatively-charged above pH 5. The pK$_a$s of the imidazole group of a free histidine and the N-terminal α-amino group are about 6 and these groups undergo ionization at neutral pH. Within a folded protein, the pK$_a$s of these ionizable groups depend on their environments and can be different from those of free amino acids or in an unfolded protein. The net charge of a protein is a function of the pK$_a$ of each ionizable group and the pH of the solution. Proteins are more positively-charged at lower pH than at higher pH.

Ion-exchange chromatography resins have either positively- or negatively-charged functional groups attached to a matrix. The diethylaminoethyl (DEAE) and quaternary aminoethyl (QAE) groups are the most commonly used functional groups for anion-exchange chromatography whereas the carboxymethyl (CM), sulfopropyl (SP), and sulfonate (S) groups are used commonly as cation-exchangers. At a given pH, positively-charged proteins will bind to cation-exchange columns and can be eluted either by increasing the ionic strength (salt concentration) or by raising the pH. Negatively-charged proteins will bind to anion-exchangers and can be eluted by increasing the ionic strength or by lowering the pH. A variety of anion- and cation-exchange materials is commercially available and the reader is referred to the manufacturers' instructions concerning properties and usage.

Alternatively, fast protein liquid chromatography (FPLC) has been designed to achieve high resolution chromatographic separation of macromolecules. Mono-Q and Mono-S columns are anion- and cation-exchangers, respectively. Fine ion-exchange materials are packed under high pressure and these columns allow high flow rates and good resolution when used at relatively high pressures. In order to obtain maximum separation, these columns should not be overloaded. Since each run normally takes less than an hour the system can be run repeatedly.

The *E. coli* cell extract is a complex mixture of charged molecules. If the recombinant protein is strongly basic (pI > 9), the cell extract can be applied directly to cation-exchange chromatographic supports such as CM– or SP–Sepharose. Negatively-charged RNA and DNA, and the majority of *E. coli* proteins will go straight through the column whereas the over-produced basic protein will bind. The target protein can then be eluted with a linear salt gradient after washing the column extensively with starting buffer.

If the over-produced protein is moderately basic, the extract should be first diluted with cold distilled water until the conductivity drops below 0.6 mMHO and then the pH adjusted to 6.0 with 3 M phosphoric acid or 1 M acetic acid (the conductivity of a 10 mM sodium phosphate solution at pH 6.0 is about 0.4 mMHO). Under these conditions the protein will bind to CM or SP supports more readily. Sometimes even a strongly basic protein, however, fails to bind to CM– or S–Sepharose. In such cases, proteins are strongly interacting with nucleic acids, which represent about one quarter of the total cell dry weight, and the bulk of nucleic acid should be removed first (10) (see below).

If the over-produced protein is negatively-charged then it will bind only to positively-charged chromatographic media such as QAE– and DEAE–Sepharose. The cell extract (pH and conductivity adjusted to those of a 50 mM Tris–HCl solution, pH 8.0) can be crudely fractionated on a DEAE–cellulose column equilibrated with 50 mM Tris–HCl pH 8.0. Most proteins can be eluted with buffer containing 0.2 M NaCl whereas the remaining DNA and RNA are eluted more slowly.

Removal of nucleic acids can be achieved by precipitation with streptomycin (11), protamine sulfate (12), or polyethyleneimine (13). Protamine is a highly basic protein purified from salmon sperm. It binds to nucleic acids, displacing some nucleic acid binding proteins. Protamine sulfate is first dissolved in a buffer of neutral pH to a concentration of 10 mg/ml and the solution is added dropwise to the cell extract until no increase in turbidity is observed. Nucleic acids are then removed by centrifugation (12). Polyethyleneimine is a positively-charged polymer which also binds to nucleic acids and precipitates them. Polyethyleneimine is first diluted to 10% and the pH is adjusted to 8.0 with HCl. The polyethyleneimine solution is added dropwise to the cell extract to 0.4% and nucleic acid precipitates removed by centrifugation. As polyethyleneimine is a positively-charged polymer, some acidic proteins also bind to it and are co-precipitated with nucleic acids. Nucleic acid binding proteins tightly bound to nucleic acids are also precipitated by polyethyleneimine. For example, λ phage repressor was precipitated by polyethyleneimine and re-dissolved in high salt buffer (14).

If the degree of purification from total cell extract by ion-exchange chromatography is unsatisfactory, the *E. coli* cell extract may be first fractionated by ammonium sulfate. Ammonium sulfate is added stepwise until the over-produced protein just begins to precipitate. The pH should be monitored constantly and re-adjusted with a 3 M Tris–HCl buffer (pH 8.5) when necessary. After centrifugation, ammonium sulfate is added to the supernatant until most of the over-produced protein is precipitated and the precipitate is collected by centrifugation. The protein pellet is re-dissolved in a minimal volume of cold, low ionic strength buffer and dialysed extensively against cold buffer to be used in the next chromatographic step.

3.3.2 Gel filtration

Gel filtration is based on the relative size of protein molecules and is performed using porous beads as the chromatographic support. A column filled with this material will have two measurable liquid volumes, the external (or void) volume between the beads, and the internal volume within the pores. A mixture of proteins is applied at the top of a gel filtration column and allowed to percolate through the column. The large protein molecules are excluded from the internal volume and therefore emerge first from the column. The smaller protein molecules which have access to the internal volume appear later.

The resolution of gel filtration is limited because none of the proteins is retained by the column during chromatography. Therefore, it should be used relatively late in a purification procedure when the number of contaminating proteins is small. For example, pooled fractions obtained from ion-exchange chromatography will contain a mixture of proteins with about the same net charge but a range of molecular weights.

3.3.3 Affinity chromatography

Affinity chromatography is one of the most powerful procedures that can be applied to protein purification. It takes advantage of the biological properties of the molecule to be purified. The target protein recognizes a specific ligand coupled to a solid support while contaminating proteins do not bind to the column.

The concept of affinity chromatography has recently been exploited also for the purification of proteins that do not have an intrinsic specific affinity for a particular ligand. By fusing an affinity tag to the target protein, the specific interaction between the tag and its ligand can be used to purify the fusion protein. Section 5 describes in more detail purification procedures using affinity handles.

4. Purification of proteins from inclusion bodies

4.1 Isolation of inclusion bodies from total cell extracts

The over-production of proteins in *E. coli* often leads to inclusion body formation. This is due to aggregation of partially folded proteins which accumulate rapidly in the cytoplasm of *E. coli*. Inclusion body formation is not restricted to foreign proteins and even endogenous soluble proteins may form inclusion bodies when over-produced. Pure inclusion bodies can be prepared by low speed centrifugation after cell lysis. It is important to lyse the cells as completely as possible to avoid contaminations with other cellular components. The low speed pellet also contains part of the membrane fraction. Triton X-100 washes remove most of the membrane proteins, including OmpA (15, 16). Suitable procedures are described in *Protocol 3*.

Protocol 3. Isolation of inclusion bodies (15)

Reagents

- Lysis buffer: 50 mM Tris–HCl pH 8.0, 25% sucrose (w/v), 1 mM EDTA
- Detergent buffer: 0.2 M NaCl, 1% deoxycholic acid (w/v), 1% Nonidet P-40 (v/v)
- Triton X-100/EDTA solution: 1% Triton X-100 (v/v), 1 mM EDTA
- DNase I (Sigma, D5025) dissolved in 0.2 M NaCl
- 1 M MgCl$_2$
- 1 M MnCl$_2$
- Lysozyme

67

Protocol 3. *Continued*

Method

1. Harvest *E. coli* cells by low speed centrifugation. Suspend the cells in lysis buffer (at least 1 ml of buffer/1 g of wet cells).

2. Dissolve lysozyme in a small volume of lysis buffer and add to the cell suspension to a final concentration of 0.1 mg/ml.

3. Incubate the cell suspension on ice for 30 min and then freeze at $-20\,°C$. Thaw the suspension by immersing the tube in water. As the solution is thawed it becomes viscous.

4. Add $MgCl_2$, $MnCl_2$, and DNase I to the suspension to final concentrations of 10 mM, 1 mM, and 10 $\mu g/ml$, respectively. The viscosity of the solution decreases as DNA is digested by DNase I. An indication of how well the DNA has been digested is obtained by pouring the solution from one tube to another. Once the solution is fluid rather than gelatinous the digestion is sufficiently complete.

5. Add an equal volume of detergent buffer to the lysate and centrifuge at 5000 *g* for 10 min. Transfer the supernatant carefully to another tube (not by decantation) without disturbing the white tight pellet and upper jelly-like layer.

6. Resuspend the pellet in the Triton X-100/EDTA solution and spin down the inclusion bodies. Repeat this procedure until the jelly-like layer is no longer seen.

4.2 Solubilization of inclusion bodies

Inclusion bodies are soluble only in strong denaturing solutions such as urea, guanidine–HCl, strong base, and acetonitrile. If an over-produced protein is soluble in urea, it can be purified readily by ion-exchange chromatography in the presence of 7–8 M urea. Cyanate is slowly formed from urea and carbamylates uncharged amino groups of proteins. Hence prolonged exposure of the protein to urea must be avoided at high pH.

While some aggregated proteins are soluble in 8 M urea (more readily at higher pH), others are often insoluble and can be pelleted by low speed centrifugation. In such cases, proteins cannot be purified by ion-exchange chromatography. This often happens when the recombinant protein is strongly basic. For instance, inclusion bodies containing human β-globin readily dissolve in 8 M urea but the more positively-charged α-globin remains insoluble and can be solubilized only in 5 M guanidine–HCl. However, when the guanidine/α-globin solution is dialysed first against water and then against 8 M urea buffer, α-globin remains soluble (17). A test for solubility is given in *Protocol 4*.

Protocol 4. Solubility test

Equipment and reagents

- Buffer containing 8 M urea
- SDS sample buffer: 25 ml 0.5 M Tris–phosphate buffer pH 6.8, 20 ml glycerol, 40 ml 10% SDS, 14 ml distilled water, and 0.1 g bromophenol blue
- SDS–polyacrylamide gel
- SDS–PAGE gel electrophoresis apparatus

Method

1. Solubilize (suspend) a small amount of inclusion bodies in buffer containing 8 M urea. Keep 100 μl of the suspension aside for the analysis by SDS–PAGE.

2. Centrifuge 1 ml of the suspension for 10 min at 80 000 r.p.m. (250 000 g) using a Beckman TL-100 table-top ultracentrifuge or equivalent. Transfer the supernatant to another tube. Resuspend the pellet in 1 ml of urea-containing buffer.

3. Separately mix the 100 μl suspension from step 1, supernatant from step 2, and the re-dissolved final pellet with equal volumes of SDS sample buffer, and analyse by SDS–PAGE (83). If the protein is found in the supernatant, it is soluble in urea and can be purified by ion-exchange chromatography in the presence of urea.

Protocol 5. Solubilization of inclusion bodies in urea

Equipment and reagents

- Sorvall SS34 rotor or equivalent
- Buffer containing 8 M urea
- Reagents and apparatus for SDS–PAGE

Method

1. Solubilize inclusion bodies in at least 10 vol. of 8 M urea buffer. Dissolve the inclusion body pellet by using a pipette or glass homogenizer.

2. Spin the solution at 20 000 r.p.m. (48 000 g) using a Sorvall SS34 rotor or equivalent. A large jelly-like pellet is normally obtained. Collect the supernatant.

3. Resuspend the pellet in urea-containing buffer and spin again at 20 000 r.p.m. (48 000 g). Collect the supernatant.

4. Analyse the supernatants from steps 2 and 3 by SDS–PAGE. Retain the pooled supernatants for further purification.

An example for the purification of inclusion bodies insoluble in urea has been described by Pavletich and Pabo (18) and is given in *Protocol 6*. Inclusion bodies containing the zinc finger peptide Zif268 were purified by the use of a reverse-phase resin.

Protocol 6. Purification of the Zif268 zinc finger peptide (18)

Equipment and reagents

- C4 reverse-phase HPLC column (Vydac) equilibrated in 0.1% TFA (trifluoroacetic acid)
- C4 resin
- 100% acetonitrile/0.1% TFA
- Water/0.1% TFA
- 6.4 M guanidine–HCl, 50 mM Tris–HCl pH 7.4

Method

1. Suspend 50 ml dry volume of C4 resin in 2–3 vol. of 100% acetonitrile/0.1% TFA. Spin down the resin and discard the supernatant. Repeat this step three times.

2. Suspend the resin in 2–3 vol. of water/0.1% TFA and spin down the resin. Discard the supernatant and repeat this step three times.

3. Dissolve inclusion bodies from 1 litre of culture in 40 ml of 6.4 M guanidine–HCl, 50 mM Tris–HCl pH 7.4. Remove any insoluble material by centrifugation.

4. Mix the guanidine extract with the C4 resin and wash the resin three times with water/0.1% TFA.

5. Elute the protein stepwise by increasing the concentration of acetonitrile/0.1% TFA.

6. Further purify the peptide on a C4 reverse-phase (Vydac) HPLC column in 0.1% TFA with an acetonitrile gradient.

4.3 Purification of denatured proteins

If the protein is soluble in urea-containing buffer, purification can be performed by ion-exchange chromatography in the presence of urea. The pK_as of Arg, Lys, Glu, Asp, and His in unfolded proteins are very similar to those of free amino acids and the net charge of a protein at a given pH can be estimated. If the total number of positively-charged residues (Arg, Lys, and His) is larger than the total number of negatively-charged residues (Glu and Asp), the protein is likely to bind to SP– or CM–Sepharose at low pH. A suitable buffer for this purification is 8 M urea/50 mM sodium acetate with the sodium acetate used from a stock solution of pH 4.0–4.5. The protein can be eluted by a linear salt gradient.

If the total number of negatively-charged amino acids is larger than the total number of positively-charged amino acids, the protein is likely to bind to QAE– or DEAE–Sepharose.

5. Use of affinity handles for purification of proteins

Recombinant DNA technology has allowed *in vitro* fusions of genes or gene fragments in a simple and predictable manner. Gene fusions were first described for heterologous bacterial expression of small peptides, such as somatostatin and insulin, combined with β-galactosidase. Since then, fusion proteins have been used to overcome protease degradation problems, to allow secretion of gene products through the cytoplasmic membrane, to generate antigens for antibody production, and to facilitate protein purification. An excellent review on gene fusion strategies has been written recently (19).

This section will focus on the affinity purification of soluble fusion proteins. The basic concept is based on the specific interaction of an additional polypeptide tag fused to the target protein with an immobilized ligand. A cell lysate including the fusion protein with the affinity tag is passed through an affinity column containing a ligand that specifically interacts with the affinity handle. The fusion protein is retained by the ligand while the other contaminating proteins can be washed through the column. After elution of the fusion protein, a chemical or enzymatic method is used to cleave the fusion protein at the junction between the two protein moieties. The cleavage mixture is again passed through the column to allow the affinity handle to bind and the target protein is collected in the flow-through fraction. The column is regenerated by elution of the bound material. Alternatively, the cleavage products (and the specific protease) can be separated by other chromatographic procedures.

Table 1 provides an overview of gene fusion systems used to facilitate protein purification. Some systems such as the glutathione S-transferase and maltose binding protein fusions require a correctly folded affinity tag, excluding purification under denaturing conditions. In contrast, poly-histidine tagged fusion proteins can be recovered under native and denaturing conditions. Many fusion tails do not interfere with the biological activity of the target protein. Several recent reviews cover multiple aspects of the fusion tail technology (20–22).

5.1 Metal chelate affinity chromatography

Immobilized metal affinity chromatography (IMAC), also known as metal chelate affinity chromatography (MCAC), is based on the co-ordination of chelated metal ions by accessible amino acid residues such as histidines in the target protein. This section will describe the principles of MCAC and will give a basic protocol for the purification of soluble target proteins suitable for

Table 1. Examples of fusion systems used to facilitate purification of soluble proteins[a]

Purification tag	Size	Ligand	Elution	Reference
Enzymes				
β-Galactosidase	116 kd	APTG	Borate pH 10	23, 24
GST	26 kd	Glutathione	Glutathione	25
CAT	24 kd	Chloramphenicol	Chloramphenicol	26, 27
PhoS	36 kd		Immunoprecipitation	28
Polypeptide binding proteins				
SPA	31 kd	IgG	pH 3.5	29, 30
ZZ	14 kd	IgG	pH 3.4	31
SPG	30 kd	Albumin	pH 2.8	32
Carbohydrate binding proteins				
MBP	40 kd	Amylose	Maltose	33–36
Antigenic epitopes				
RecA	144 or 329 aa	Anti-RecA antibody	pH 2.5	37
c-myc	11 aa	9E10	pH 3.0	38
FLAG	8 aa	Anti-FLAG M1	EDTA pH 7.4	39, 40
		Anti-FLAG M2	pH 3.5	41
Poly(amino acid) tails				
Poly(Arg)	5 aa	Ion-exchange	Salt gradient	42
Poly(Asp)	5–16 aa	Ion-exchange	Salt gradient	43, 44
		Precipitation	Polyethyleneimine	
Glu	1 aa	Ion-exchange		45
Poly(His)	2–6 aa	MCAC	Low pH	46
			Imidazole pH 7.4	47
			EDTA	
Poly(Cys)	4 aa	Thiopropyl	Cysteine and dithiothreitol	48
Poly(Phe)	11 aa	Phenyl	Ethylene glycol	48

[a] A few references for each fusion protein are given. Abbreviations used: aa, amino acid residues; APTG, p-aminophenyl-β-D-thiogalactopyranoside; CAT, chloramphenicol acetyltransferase; GST, glutathione S-transferase; IgG, immunoglobulin G; MBP, maltose binding protein; MCAC, metal chelate affinity chromatography; SPA, Staphylococcal protein A; SPG, Streptococcal protein G; ZZ, synthetic protein A analogue.

MCAC. In addition, further information will be provided for eventual problems arising during the purification procedure. The reader is also referred to the instructions supplied by the manufacturers of chelating resins.

Metal affinity separations exploit the affinities for metal ions that are exhibited by functional groups on the surface of proteins (46). Although a number of functional groups participate in metal binding in metalloproteins, the actual situation for MCAC is less complex: the side-chains of histidine dominate protein binding to chelated transition metal ions, for example Cu^{2+}, Ni^{2+}, or Zn^{2+}. Other side-chains and functional groups make much smaller contributions to a protein's apparent affinity for the chelated metal. Proteins are retained on metal affinity columns according to the number of accessible surface histidines. However, individual histidyl residues vary in their affinities for immobilized metal ions, depending on the influence of neighbouring residues. Histidine is a relatively rare amino acid, accounting for about 2% of the amino acids in globular proteins. Only about half of the histidine residues are exposed on the protein surface. Proteins that contain neighbouring histidines are not common in bacteria. Therefore, target proteins with genetically introduced poly-histidine tails display a high affinity interaction with metal ions and permit efficient purification. The strength of interaction between a histidine-tailed protein and the affinity matrix is dependent on many factors such as length of the histidine tail, the choice of the metal ion, the pH, and the properties of the target protein itself (see below).

In most cases, a poly-histidine tail (located either at the N- or C-terminus) does not interfere with the biological activity of the target protein. Therefore, there is often no need to remove the histidine tail after purification of the target protein.

In order to utilize the protein–metal ion interaction for chromatographic purposes, the metal ion must be immobilized on to an insoluble support. This is achieved by attaching a chelating group to the chromatographic matrix. One chelating group used in this technique is iminodiacetic acid (IDA), coupled to a matrix via a long hydrophilic spacer arm. The spacer arm ensures that the chelating metal (for example Cu^{2+} or Zn^{2+}) is fully accessible to all available binding sites on a protein. IDA is a tridentate chelator. As the metal ions will co-ordinate four to six ligands, the remaining co-ordination sites are occupied by water molecules or buffer components, which can be displaced by appropriate protein functional groups. Nitrilotriacetic acid (NTA) with four chelating sites has been used in combination with Ni^{2+} (47, 49).

The most commonly used metals for MCAC are Cu^{2+}, Ni^{2+}, or Zn^{2+}. Protein retention on different metals reflects the affinity of the metal ion for imidazole: both protein retention and stability constants for complexation with imidazole follow the order $Cu^{2+} > Ni^{2+} > Zn^{2+}$ (50). The choice of the best metal is not always predictable. Copper often affords much tighter binding to proteins than does zinc. However, the weaker binding achieved using zinc may be exploited for selective elution of a protein mixture in some cases. Therefore, the appropriate choice of the metal ion may have to be found in a trial and error process.

The purification of histidine-tailed fusion proteins is outlined in *Protocols 7* and *8*. If the fusion protein is to be purified from the cytoplasm, use *Protocol 7*. If the fusion protein is to be purified from the periplasm, use *Protocol 8*.

Protocol 7. Purification of recombinant cytoplasmic proteins by
MCAC under native conditions

Equipment and reagents

- Buffers for preparation of total cell lysate:
  Wash buffer A: 50 mM sodium phosphate
  pH 8.0
  Lysis buffer[a]: 50 mM sodium phosphate
  pH 8.0, 500 mM NaCl
- Wash buffer B: 50 mM sodium phosphate
  pH 6.0, 500 mM NaCl

- Chelating resin: Chelating Sepharose Fast
  Flow (Pharmacia, Sigma) charged with
  Zn^{2+} (capacity: 20–30 μmole Zn^{2+}/ml gel)
  or NTA resin (QIAGEN) loaded with Ni^{2+}
  (capacity: 8–12 μmole Ni^{2+}/ml or 5–10 mg
  of histidine-tagged protein/ml)
- 2.5 × 10 cm chromatography column

Method

1. Prepare total cell extract from a 1 litre bacterial culture by sonication in
   25 ml ice-cold lysis buffer (see *Protocol 1*, steps 1–4).

2. Centrifuge for 10 min at 8000 g and 4°C to remove insoluble material
   and intact cells. Carefully collect supernatant only.

3. Pour chelating resin loaded with metal ions into a 2.5 × 10 cm column.
   Wash with 8 vol. of ice-cold lysis buffer. The amount of resin depends
   on the amount of fusion protein produced. The capacities of the respective resins are given under *Equipment and reagents*.

4. Load the column with the crude cell extract at a flow rate of about 1 ml/
   min.

5. Wash the column with ice-cold lysis buffer until the A_{280} of the flow-
   through is less than 0.01.

6. Wash with ice-cold wash buffer B until the A_{280} of the flow-through is
   less than 0.01.[b]

7. Elute the protein with a gradient of 0–500 mM imidazole in wash buffer
   B at pH 6.0. Collect 1 ml fractions. Alternatively, elute the protein with a
   pH 6.0 to pH 4.0 gradient in wash buffer B.

---

[a] Chelating agents such as EDTA or citrate should not be included. 2-Mercaptoethanol does not
interfere with the purification procedure, but stronger reducing agents such as dithiothreitol
should be avoided because the metal ions will be reduced.
[b] This wash step is optional and requires that the histidine-tagged target protein remains bound
on the column resin at pH 6.0. Contaminating proteins may elute under these conditions.

Protocol 8. Purification of recombinant periplasmic proteins by
MCAC under native conditions

Equipment and reagents

- Lysis buffer, chelating resin, chromatography column, wash buffer B (see
  Protocol 7)

- Buffers for the preparation of the periplasmic protein fraction (*Protocol 2*)

Method

1. Prepare periplasmic protein fraction from a 1 litre bacterial culture by cold osmotic shock (*Protocol 2*).

2. Dialyse cold osmotic shock fluid extensively against lysis buffer[a] before continuing with the purification to remove EDTA.

3. Pour chelating resin loaded with metal ions into a 2.5 × 10 cm chromatography column. Wash with 8 vol. of ice-cold lysis buffer. The amount of resin depends on the amount of fusion protein produced. The capacities of the respective resins are given in *Protocol 7, Equipment and reagents*.

4. Load the column with the periplasmic protein fraction at a flow rate of about 1 ml/min.

5. Wash the column with ice-cold lysis buffer until the A_{280} of the flow-through is less than 0.01

6. Wash the column with ice-cold wash buffer B until the A_{280} of the flow-through is less than 0.01.[b]

7. Elute the protein with a gradient of 0–500 mM imidazole in wash buffer B at pH 6.0. Collect 1 ml fractions. Alternatively, elute the protein with a pH 6.0 to pH 4.0 gradient in wash buffer B.

[a] Chelating agents such as EDTA or citrate should not be included.
[b] This wash step is optional and requires that the histidine-tagged target protein will remain bound on the column resin at pH 6.0. Contaminating proteins may elute under these conditions.

Binding of the target protein usually occurs in the pH range 7 to 8. As the pK_as of surface histidyl residues are generally between 6 and 7, the imidazole nitrogen will be in the unprotonated state co-ordinating the metal ion. The choice of starting buffer depends on the chelated metal and on the binding properties of the sample molecules. Sodium acetate, sodium phosphate, and Tris–acetate are suitable buffers. Tris–HCl tends to reduce binding and should be used when the metal protein affinity is fairly high. Chelating agents such as EDTA or citrate should not be included. 2-Mercaptoethanol (up to 10 mM) does not interfere with the purification procedure. However, stronger reducing agents such as dithiothreitol should be avoided because the metal ions will be reduced.

The use of high concentrations of salt or detergent in buffers normally does not affect the adsorption of the protein to the metal column. Incorporation of sodium chloride (0.5–1 M) in all buffers used will eliminate ion-exchange effects.

The target protein can be desorbed from the affinity matrix by three main procedures:

(a) A pH gradient. Since the histidine imidazole nitrogen co-ordinates metal ions in the unprotonated state, a decrease in pH is sufficient to elute proteins. As several proteins may bind to the gel, optimal purification is achieved with a decreasing pH gradient, normally in the range between pH 7 to pH 4.

(b) A competitive ligand. Imidazole competes with the protein ligands for the metal ions. Good separations are obtained by eluting with an increasing concentration gradient of imidazole (0–500 mM).

(c) A chelating agent. Chelating agents such as EDTA or EGTA will strip the metal ions from the gel and cause the elution of all adsorbed proteins. However, this method does not resolve different proteins.

The individual binding properties of histidine-tailed target proteins allow the application of a wide variety of elution conditions. Lowering the pH to 6.0 in a wash step before elution may eliminate contamination, but leave the target protein bound. If it is desirable to keep the pH above pH 7.0 at all times, the column can be washed with buffer containing up to 40 mM imidazole before application of the competitive ligand gradient. For optimizing elution conditions, the imidazole and pH gradient may be combined. It is up to the researcher to establish optimal purification conditions for the respective target protein.

The binding of histidine-tagged proteins to the resin does not require any functional protein structure and is thus unaffected by strong denaturants such as guanidine hydrochloride or urea. This allows protein also to be efficiently purified from solubilized inclusion bodies.

Examples of purification of histidine-tailed proteins by MCAC are given in *Table 2*.

5.2 Glutathione S-transferase fusion proteins

The glutathione S-transferase (GST) system is designed for intracellular expression of genes or gene fragments as fusions with glutathione S-transferase

---

Table 2. Examples of soluble fusion proteins purified by MCAC[a]

Target protein	Tail	Ligand	Elution	Reference
Fv antibody fragment	$(His)_5$ (C)	Zn^{2+}, IDA	Imidazole	51
HIV-1 reverse transcriptase	$(His)_6$ (N)	Ni^{2+}, NTA	pH gradient	52
Mouse dihydrofolate reductase	$(His)_2$ (N,C)	Ni^{2+}, NTA	pH gradient	47
Mouse dihydrofolate reductase	$(His)_6$ (N,C)	Ni^{2+}, NTA	pH gradient in guanidine–hydrochloride	47
Transcription factor USF	$(His)_6$ (N)	Ni^{2+}, IDA	Imidazole	53

[a] Abbreviations used: C, histidine tail at C-terminus; IDA, iminodiacetic acid; N, histidine tail at N-terminus; NTA, nitrilotriacetic acid.

encoded by the parasitic helminth *Schistosoma japonicum*. In the majority of cases, fusion proteins are soluble and can be purified from crude bacterial lysates under mild, non-denaturing conditions by affinity chromatography on immobilized glutathione (25). The C-terminal extension of the 26 kd GST carrier by the target protein does not affect the binding to immobilized glutathione. This is consistent with a model in which the carrier and the target protein form independent domains. Recognition sites for specific proteases such as thrombin or blood coagulation factor Xa (encoded by the respective expression vectors) between the GST moiety and the target protein allow the proteolytic cleavage of the desired polypeptide from the fusion product.

The purification of GST fusion proteins by a batch procedure is outlined in *Protocol 9*. The addition of Triton X-100 to the cell extract minimizes the association of the fusion protein with bacterial proteins, and thus prevents the appearance of contamination in the final preparation. A single affinity chromatography step can generate fusion protein preparations that are more than 90% pure.

The purification conditions described in *Protocol 9* should be regarded as initial guide-lines. The amount of glutathione agarose needed is dependent on the expression levels of the recombinant GST fusion protein. If the fusion protein is not quantitatively eluted, the glutathione concentration in the elution buffer may be increased to 10 mM or 15 mM (54) and/or salt added. The overall yield of fusion protein can sometimes be improved by increasing the quantity of glutathione agarose beads, minimizing the volume of liquid during adsorption, and extending the period of adsorption.

Alternatively to the batch procedure of *Protocol 9*, the purification can be performed using column chromatography equipment. For detailed information, the reader is referred to the instructions supplied by the manufacturers of immobilized glutathione materials.

Protocol 9. Purification of GST fusion proteins by affinity chromatography

Equipment and reagents

- Buffers for preparation of total cell lysate: Wash buffer (PBS): 150 mM NaCl, 16 mM Na_2HPO_4, 4 mM NaH_2PO_4, pH 7.3 Lysis buffer: PBS
- Elution buffer: 5 mM reduced glutathione in 50 mM Tris–HCl pH 8.0, freshly prepared, final pH 7.5
- S-linked glutathione agarose beads (Sigma or Pharmacia). Pre-swell the beads in 10 vol. PBS for 1 h at room temperature. Wash twice with PBS and store as a 50% slurry at 4°C in PBS with 1 mM sodium azide. The capacity is > 5 mg glutathione S-transferase/ml drained gel.
- Bench-top centrifuge holding 50 ml tubes
- Microcentrifuge

Method

1. Prepare total cell extract from a 1 litre bacterial culture by sonication in 25 ml ice-cold lysis buffer (see *Protocol 1*, steps 1–4).

Protocol 9. *Continued*

2. Add Triton X-100 at a final concentration of 1% (v/v) and mix gently.

3. Centrifuge 10 min at 8000 *g* and 4 °C to remove insoluble material and intact cells. Carefully collect the supernatant only.

4. Add 2 ml of the 50% slurry of glutathione agarose beads to the supernatant and mix gently for 5 min.

5. Collect the beads by centrifugation for 10 sec at 500 *g*.

6. Wash the beads by adding 50 ml ice-cold PBS, mixing, and centrifuging 10 sec at 500 *g*. Repeat the wash two more times.

7. Resuspend the beads in 1 ml of ice-cold PBS and transfer to a narrow tube.

8. Centrifuge 10 sec at 500 *g* to collect the beads and discard the supernatant.

9. Elute the fusion protein by adding 1 ml elution buffer. Mix gently for 2 min, centrifuge 10 sec at 500 *g*, and collect the supernatant. Repeat the elution two more times and analyse each fraction by SDS–PAGE.

The GST system requires an active, correctly folded GST moiety. Therefore, the purification of fusion proteins under denaturing condition is not possible. However, the binding of GST fusion proteins to glutathione beads is not disrupted by the presence of 1% Triton X-100, 1% Tween-20, 10 mM dithiothreitol, or 0.03% SDS (25). This offers the possibility (to a certain degree) to purify insoluble GST fusion proteins by affinity chromatography.

Removal of the GST moiety from the fusion protein is achieved by cleavage with a site-specific protease (dependent on the expression vector used). The processing of fusion proteins is discussed in Section 6.

Examples of purification of GST fusion proteins are given in *Table 3*.

Table 3. Examples of GST fusion proteins[a]

Target protein	Elution	Reference
Human Pim-1 proto-oncogene product	10 mM GSH, pH 8.0	55
ATIC	5 mM GSH, pH 8.0	56
IgG binding domain of SPA	2 mg/ml GSH, pH 9.6	57
IgG binding domain of SPG	2 mg/ml GSH, pH 9.6	57
E1A binding domain of retinoblastoma gene product	SDS	58
Outer surface protein A from *Borrelia burgdorferi*	5 mM GSH, pH 7.3	59

[a] Abbreviations used: ATIC, 5-aminoimidazole-4-carboxamide-ribonucleotide transformylase-inosine monophosphate cyclohydrolase; GSH, reduced glutathione; GST, glutathione S-transferase; SPA, Staphylococcal protein A; SPG, Streptococcal protein G.

5.3 Maltose binding protein fusions

Several fusion protein recovery systems have been designed to take advantage of low cost complex carbohydrates as adsorbents, utilizing mild conditions for both binding and elution (20). The periplasmic maltose binding protein (MBP) of *E. coli* has been used as an N-terminal tail for purification of target proteins either expressed intracellularly or secreted to the periplasm. Expression plasmids are commercially available (New England Biolabs) (33–36) encoding MBP with or without its signal peptide. A polylinker located 3' of the *malE* gene allows in-frame insertion of the target gene. The MBP vectors also include a sequence that codes for the factor Xa cleavage site, placed so that it can be used to separate the target protein from the MBP moiety after affinity purification.

Purification of the MBP fusion protein is based on the strong affinity of MBP (370 residues, 40 kd) to cross-linked amylose. After binding, MBP fusion proteins can be eluted under very mild, physiological conditions with maltose. This enables the target protein to keep its biological activity. The efficiency of the chromatographic purification is over 70%.

The purification of MBP fusion proteins is outlined in *Protocols 10* and *11*. If the fusion protein is to be purified from the cytoplasm, use *Protocol 10*. If the fusion protein is to be purified from the periplasm, use *Protocol 11*. The affinity chromatography step should be carried out in one day. If the MBP fusion protein is left on the column too long, it starts dissociating from the immobilized amylose, possibly due to maltose that is released from the column matrix by trace amounts of amylase co-purified from *E. coli* along with the fusion protein (54).

Protocol 10. Purification of maltose binding protein fusions from total cell extract by affinity chromatography

Equipment and reagents

- Buffers for preparation of total cell lysate:
  Wash buffer: 50 mM Tris–HCl pH 7.4
  Lysis buffer: 20 mM Tris–HCl pH 7.4, 200 mM NaCl, 10 mM EDTA, 10 mM 2-mercaptoethanol
- Column buffer A: 20 mM Tris–HCl pH 7.4, 200 mM NaCl, 1 mM EDTA, 10 mM 2-mercaptoethanol
- Amylose resin (New England Biolabs): Pre-

swell by adding 30 ml column buffer A per gram resin and incubating at room temperature for 30 min. The bed volume is about 15 ml/g resin. The binding capacity is 2–3 mg fusion protein/ml bed volume.
- Elution buffer: column buffer A containing 10 mM maltose
- 2.5 × 10 cm chromatography column

Method

1. Prepare total cell extract from a 1 litre bacterial culture by sonication in 25 ml ice-cold lysis buffer (see *Protocol 1*, steps 1–4).

2. Centrifuge 10 min at 8000 *g* and 4°C to remove insoluble material and intact cells. Carefully collect the supernatant only.

Protocol 10. *Continued*

3. Pour amylose resin into a 2.5 × 10 cm column. Wash with 8 vol. of ice-cold column buffer A. The amount of resin required depends on the amount of fusion protein produced.

4. Dilute the crude cell extract 1:10 with ice-cold column buffer A.[a] Load this on to the column at a flow rate of about 1 ml/min.

5. Wash the column with 8 vol. of ice-cold column buffer A.

6. Elute the fusion protein with ice-cold elution buffer. Collect the eluate in 3 ml fractions.

7. Pool those fractions containing the fusion protein. If necessary, concentrate to about 1 mg/ml in a Centricon or Centriprep concentrator or equivalent.

[a] The dilution of the crude extract is aimed at reducing the total protein concentration to about 2.5 mg/ml.

Protocol 11. Purification of maltose binding protein fusions from the periplasm by affinity chromatography

Equipment and reagents

- Buffers for the preparation of the periplasmic protein fraction (see *Protocol 2*)
- Column buffer B: 20 mM Tris–HCl pH 7.4, 200 mM NaCl, 1 mM EDTA
- Amylose resin (New England Biolabs). Pre-swell by adding 30 ml column buffer B per gram resin and incubating at room temperature for 30 min. The bed volume is about 15 ml/g resin. The binding capacity is 2–3 mg fusion protein/ml bed volume.
- Elution buffer: column buffer B containing 10 mM maltose
- 2.5 × 10 cm chromatography column

Method

1. Prepare the periplasmic protein fraction from a 1 litre bacterial culture by cold osmotic shock (see *Protocol 2*).

2. Add 2 ml of 1 M Tris–HCl pH 7.4, 4 ml of 5 M NaCl, and 5 ml of 100 mM EDTA pH 7.4 solution to 100 ml cold osmotic shock fluid. Increase the volume to 250 ml with ice-cold column buffer B.

3. Pour amylose resin into a 2.5 × 10 cm column. Wash with 8 vol. of ice-cold column buffer B. The amount of resin required depends on the amount of fusion protein produced.

4. Load the protein sample on to the column at a flow rate of about 1 ml/min.

5. Wash the column with 8 vol. of ice-cold column buffer B.

6. Elute the fusion protein with ice-cold elution buffer. Collect the eluate in 3 ml fractions.

7. Pool those fractions containing the fusion protein. If necessary, concentrate to about 1 mg/ml in a Centricon or Centriprep concentrator or equivalent.

Many of the MBP fusion proteins made to date have been successfully purified by affinity chromatography. In the cases that have not worked, the fusion protein binds to the column poorly or not at all, is degraded by *E. coli* proteases, or is insoluble. Proteolysis of the fusion protein leads to free MBP, blocking the affinity column. The use of cytoplasmic and periplasmic protease deficient *E. coli* strains as well as changed expression conditions may lead to stabilization of the fusion protein. Furthermore, use of *E. coli* with a deleted *malE* gene avoids the expression of wild-type MBP from the bacterial chromosome and its copurification with the MBP hybrids.

Removal of the MBP moiety from the fusion protein is achieved by cleavage with a site-specific protease (factor Xa). The processing of fusion proteins is discussed in Section 6.

Examples of purification of MBP fusion proteins are given in *Table 4*.

Table 4. Examples of maltose binding protein fusions[a]

Target protein	Location in *E. coli*	Reference
S. aureus nuclease A	PP	33
Klenow fragment	PP, CP	33
α-Keto acid decarboxylase E1	CP	60
FK506 binding protein	CP	61
HIV-1 protease	CP	62
BPTI	PP	63

[a] Abbreviations used: BPTI, bovine pancreatic trypsin inhibitor; CP, cytoplasm; PP, periplasm.

5.4 Fusions to Staphylococcal protein A

The immunoglobulin G (IgG) binding domains of Staphylococcal protein A (SPA) have a strong affinity for the constant region of mammalian IgGs, permitting the SPA fusion proteins to be purified by immunoaffinity chromatography. This fusion technique has been used successfully for the high level production of peptide hormones, for the immobilization of enzymes, and for the production of specific antibodies against gene products (ref. 29 and references therein). SPA is a protein bound to the cell wall of the pathogenic bacterium *Staphylococcus aureus*. It consists of three structurally and functionally different regions: an N-terminal signal peptide which is cleaved off after translocation, five highly homologous IgG binding domains (A to E, each with approximately 58 amino acids), and the C-terminal region X which anchors SPA in the cytoplasmic membrane and cell wall. Expression plasmids containing two synthetic IgG binding domains (based on domain B of SPA) or all five domains as N-terminal tails are available for either intracellular expression of fusion proteins in *E. coli* or, by including the SPA signal sequence, for export into the periplasm of *E. coli* or extracellular secretion from *S. aureus* (29, 30).

The IgG binding region of SPA has an extended tertiary structure consisting of five small globular domains. This characteristic facilitates independent folding of the target protein and SPA domains and therefore minimizes steric interference of the target protein on IgG binding. SPA fusion proteins are bound to immobilized IgG at neutral pH. Elution is achieved by lowering the pH to about 3.5. Therefore, the target protein is exposed for some time to relatively harsh conditions that may affect its biological activity. *Protocol 12* describes the detailed procedures.

The peptide hormones that have been thus far expressed and purified using SPA fusions have all been cleaved by chemical methods such as cyanogen bromide, weak acid, or hydroxylamine. More recently, expression vectors have been designed encoding recognition sequences for a specific protease (enterokinase, collagenase, factor Xa) between the SPA moiety and the target protein (29).

Examples of purification of SPA fusion proteins are given in *Table 5*.

Protocol 12. Purification of SPA fusion proteins from total cell extract by immunoaffinity chromatography

Equipment and reagents

- Buffers for preparation of total cell lysate: Wash buffer: 50 mM Tris–HCl pH 7.4 Lysis buffer: 50 mM Tris–HCl pH 7.4, 150 mM NaCl
- Column buffer: 50 mM Tris–HCl pH 7.4, 150 mM NaCl, 0.05% Tween-20 (v/v)
- Elution buffer: 0.5 M acetic acid titrated to pH 3.5 using ammonium acetate
- IgG agarose beads (Sigma) or IgG Sepharose Fast Flow (capacity: 2 mg protein A/ml drained gel) (Pharmacia)
- 2.5 × 10 cm chromatography column

Method

1. Prepare total cell extract from a 1 litre bacterial culture by sonication in 25 ml ice-cold lysis buffer[a] (see *Protocol 1*, steps 1–4).

2. Centrifuge 10 min at 8000 g and 4 °C to remove insoluble material and intact cells. Carefully collect the supernatant only.

3. Pour the IgG agarose resin, equilibrated with column buffer, into a 2.5 × 10 cm column. The amount of resin required depends on the amount of fusion protein produced (see above).

4. Load on to the column at a flow rate of about 1 ml/min.

5. Wash the column extensively with at least 10 vol. of ice-cold column buffer.

6. Wash with 5 vol. of 1 mM ice-cold ammonium acetate pH 5.0.[b]

7. Elute the fusion protein with ice-cold elution buffer.

8. Neutralize the eluate with 1 vol. of 1 M Tris–HCl pH 7.4.

[a] Avoid reducing agents since the disulfide bonds in IgG will be affected.
[b] This wash is performed to lower the buffer capacity of the gel matrix.

Table 5. Examples of protein A fusion proteins[a]

Target protein	Location in *E. coli*	Cleavage	Reference
Insulin-like growth factor I	Culture medium	Hydroxylamine	31
DNA binding domain of rat androgen receptor	CP	—	64
Tropoelastin	CP	Cyanogen bromide	65
Mutant BPTI	Culture medium, PP	—	29

[a] Abbreviations used: BPTI, bovine pancreatic trypsin inhibitor; CP, cytoplasm; PP, periplasm.

Protein A fusion technology has recently been extended to a dual affinity fusion strategy for stabilization and full-length isolation of recombinant proteins. Using this approach, the target protein is fused between two heterologous domains with specific affinity for two different ligands. The full-length protein can be selectively recovered by two subsequent affinity purification steps. This approach is suitable for expression of proteins which are highly susceptible to proteolysis. It is important that the target protein be allowed to fold into a biologically active structure without obstruction from the flanking heterologous domains. Small eukaryotic proteins such as human insulin-like growth factor II and proinsulin have been provided with a pair of synthetic SPA IgG binding domains at the N-terminus and the human serum albumin binding domain from Streptococcal protein G at the C-terminus (66, 67). Sequential affinity purification on immobilized IgG followed by purification on immobilized albumin greatly increased the recovery of full-length target protein. Furthermore, the presence of the C-terminal tail stabilized the fusion protein considerably.

5.5 β-Galactosidase fusion proteins

The protein β-galactosidase has been extensively used as a fusion tail for the recovery of proteins expressed intracellularly in *E. coli*. Fusions containing β-galactosidase often yield insoluble inclusion bodies. However, many examples are known which are soluble and allow affinity purification with substrate analogues as ligands (23, 24). Despite the versatility of the β-galactosidase system, the large size of the enzyme reduces the attractiveness of the system, as even high expression levels give rather low yields of the desired target product. The active enzyme consists of a tetramer, which might interfere with the folding of the fused target protein. It is not possible to obtain secretion using β-galactosidase as the fusion partner.

The β-galactosidase fusion proteins have been purified by immunological approaches (68) or by affinity chromatography with substrate analogues such as *p*-aminophenyl-β-D-thiogalactopyranoside (APTG) (23). The fusion protein is applied to the APTG column at neutral pH and high salt concentration (1.6 M NaCl) and eluted with sodium borate buffer, pH 10.0. The eluted fusion proteins are about 85% pure.

5.6 Epitope tagging and immunoaffinity purification

Immunoaffinity chromatography exploits the specific recognition of an antigen by immobilized antibodies. Antigenic epitopes can be introduced in any target protein by means of recombinant DNA techniques. This is of great advantage if antibodies recognizing the target protein are not available. Fusion proteins with antigenic epitopes can be detected by Western blot analysis or ELISA and purified on antibody columns.

Ideally, the epitope should be easily accessible on the surface of the fusion protein and furthermore should not interfere with the biological activity of the target protein. Small epitopes such as the c-myc tag (69) recognized by the mouse monoclonal antibody 9E10 (70) have been successfully used to purify single chain Fv antibody fragments produced in *E. coli* (38). However, elution of the tagged antibody fragments has been performed at low pH requiring a remarkable stability of the target protein.

Another extremely hydrophilic tail of only eight amino acids (39) (FLAG peptide) has been exploited for one-step purification of recombinant target proteins such as granulocyte–macrophage colony stimulating factor (GM-CSF) by affinity to a calcium-dependent immobilized monoclonal antibody (40). The tail was designed for fusion at the N-terminus of secreted proteins, since the calcium-dependent antibody recognized a free N-terminal FLAG peptide only. The fusion protein was bound to the antibody column in the presence of calcium and recovered by elution with a wash buffer lacking calcium ions. This allowed the purification of fusion proteins under very mild conditions.

Recently, antibodies recognizing a C-terminal FLAG tag have been made available (International Biotechnologies Inc.). However, these antibodies are not calcium ion-sensitive and elution of the fusion protein must be performed at low pH.

5.7 Avidin/streptavidin-based purification methods

Biotin is an excellent candidate for a purification tag because of its small size. However, two drawbacks have limited its usefulness so far. First, the very strong affinity of biotin for avidin and streptavidin requires harsh conditions for elution from an affinity resin. Secondly, *in vivo* biotination, which would allow specific labelling of a target protein in a crude cell lysate, has not been possible. Recently however, *in vivo* biotination of recombinant proteins in *E. coli* has been reported (84). This approach may turn out to be useful in future for the purification of target proteins. A biotin acceptor sequence, derived from carboxylases and fused to the target protein, is post-translationally modified at a specific lysine residue by *E. coli* biotin ligase. Since *E. coli* contains naturally only one biotinated protein, the biotin carboxyl carrier protein of acetyl-CoA carboxylase, the recombinant biotinated protein is one of only two biotinated proteins detectable in the cell lysate. High level production of some recombinant fusion proteins leads to decreased biotination of the *E. coli* biotin carboxyl carrier protein and growth inhibition. However, providing high concentrations of biotin in the growth medium and/ or co-expression of biotin ligase can overcome the growth inhibitory effects of fusion protein production while increasing the yields of biotinated fusion protein (84). Purification of biotinated proteins without denaturation can be achieved by using modified monomeric avidin with decreased affinity for biotin, thus permitting elution of biotinated proteins by competition with free biotin. The target protein can be released from the biotin acceptor moiety by specific proteolytic cleavage (for example with factor Xa) if suitable sites have been introduced between the biotin acceptor sequence and the recombinant protein. Biotination of target proteins has been used for their purification (84–86) and for screening for *in vivo* protein–protein interaction (87). An expression vector system is commercially available from Promega (PinPoint Protein Purification System) employing the biotin acceptor sequence (12.5 kd) from the *Propionibacterium shermanii* transcarboxylase 1.3S subunit.

A different strategy has been chosen by Schmidt and Skerra (88). A nine amino acid C-terminal affinity peptide (streptavidin-affinity tag or 'Strep-tag') was identified from a random library and shown to interact with streptavidin. It has been used for the detection and purification of a functional immunoglobulin Fv fragment produced in *E. coli*.

6. Processing of fusion proteins

6.1 Cleavage of fusion proteins

We have shown in the preceding sections that fusion protein expression methods are useful in increasing the expression level, and in facilitating purification. If the over-produced proteins are to be used for functional or

Table 6. Methods for chemical and enzymatic cleavage of fusion proteins

Cleavage site	Reagent	Comments	Reference
-Met- ↓ -	Cyanogen bromide	70% formic acid or 0.1 M HCl	71
- ↓ -Cys-	NTCB	pH 9.2	72
-Asp- ↓ -Pro-	Acid pH	70% formic acid, 48 h	73
-Asn- ↓ -Gly-	Hydroxylamine	pH 9.0, 45°C, 4h	74
-Arg- ↓ or -Lys- ↓	Trypsin	pH 7–9	75
-Arg- ↓	Clostripain	pH 8	76
-(Asp)$_4$-Lys- ↓ -	Enterokinase		77
-Ile-Glu-Gly-Arg- ↓ -	Factor Xa	pH 8	15
-Pro-X- ↓ -Gly-Pro-Y- ↓ -	Collagenase		73
-Phe-Ala-His-Tyr- ↓ -	H64A subtilisin	pH 8	78

structural studies, extra-polypeptide chains must be cleaved off by chemical or enzymatic methods. *Table 6* summarizes the chemical and enzymatic cleavage methods that may be useful for this purpose.

Most chemical cleavages are carried out under harsh conditions such as in strong acid where most proteins are unfolded. Great care must be taken not to modify other amino acid residues as the deamidation of asparagine and glutamine can take place at acidic pH and cysteine and methionine may undergo oxidation. The specificity of most chemical cleavage methods depends on only one amino acid; therefore they are not useful in cleaving large proteins even though methionine and cysteine are rather rare amino acids. The dipeptide sequence Asp–Pro undergoes acid cleavage whereas hydroxylamine cleavage depends on the Asn–Gly dipeptide.

Enzymatic cleavages are carried out under mild conditions where most proteins are very stable and can be used for both folded and unfolded proteins. In contrast to restriction endonucleases, the specificity of most proteolytic enzymes primarily depends on only one amino acid.

Trypsin cleaves peptide bonds preceded by Arg or Lys but the rate of cleavage substantially depends on the context and structure around the cleavage site. If a protein folds into a compact structure it is normally stable against such proteases. Varadarajan *et al.* (79) produced a fusion protein with human myoglobin containing several arginines and lysines. The extra phage polypeptide at the N-terminal end was cleaved off with trypsin leaving folded myoglobin intact. If the target protein within the fusion is folded into a compact domain, the extra sequence can be removed with enzymes such as trypsin and chymotrypsin. In this case the cleavage site should not be too close to the folded domain and should be accessible to the enzyme.

The recognition sequence for blood coagulation factor Xa comprises four amino acids, Ile–Glu–Gly–Arg (15). This enzyme is one of the proteolytic enzymes in the blood coagulation cascade and is homologous to trypsin. It activates prothrombin by cutting two peptide bonds within prothrombin, both of which are preceded by Ile–Glu–Gly–Arg. A fusion protein consisting of human β-globin and a part of the λ phage cII protein was specifically cleaved only at the junction when the Ile–Glu–Gly–Arg sequence was inserted between the cII polypeptide and human β-globin. Human β-globin, containing six arginine and 14 lysine residues, remains intact after factor Xa treatment whereas trypsin cleaves at every arginine and lysine under the same conditions. Factor Xa has been applied for cleavage of many fusion proteins and has proved to be one of the most useful sequence-specific proteases.

Enterokinase is the enzyme which activates trypsinogen to trypsin and has a strong preference for Arg preceded by a run of acidic amino acids. Collagenase has also been used to cleave fusion proteins. These three proteases (factor Xa, enterokinase, collagenase) have a strong sequence preference but should not be regarded as strict sequence-specific enzymes as some polypeptides are known to be cleaved even when the recognition sequences are not present.

To overcome this difficulty, Carter *et al.* (78) have developed a substrate assisted cleavage method. Subtilisin has a broad specificity but preferentially cleaves peptide bonds preceded by a large hydrophobic residue. His 64 of subtilisin is part of a catalytic triad and when this residue is mutated to Ala the enzyme becomes inactive. When His precedes Tyr or Phe at the P1 site within substrates, His can substitute the missing His 64 of subtilisin and the peptide bond is cleaved efficiently. This inactive enzyme can only cleave a substrate containing His–Tyr which is rare in natural proteins. This method has enormous potential for its application to fusion protein cleavage.

6.2 *In vitro* re-folding of proteins

Proteins produced in *E. coli* often form inclusion bodies and these proteins can only be solubilized with denaturing reagents such as urea and guanidine–HCl. In order to cleave fusion proteins chemically it is often necessary to expose them to strong acid, causing complete unfolding. These proteins must be re-folded to their native and functional forms in order to carry out biochemical and structural studies. It is generally believed that a protein has the lowest free energy in its native conformation and should fold into the native conformation spontaneously. Nevertheless, the *in vitro* re-folding of proteins is usually not straightforward and some proteins have not been re-folded *in vitro* despite extensive efforts. The nascent polypeptides probably start folding using short-range interactions as they emerge from the ribosome. Long-range interactions gradually build up to complete the folding process. Some people believe that *in vitro* the full-length protein could be trapped in a local free energy minimum and that re-folding is not possible unless this co-translational folding process is re-produced. Nevertheless protein chemists have made tremendous efforts to re-fold proteins, with limited success. Experimentally, proteins are taken through various solution conditions to encourage folding. This is very much a trial and error process. The major obstacle of spontaneous re-folding is aggregation of proteins. There are no general protocols for re-folding proteins but in all cases proteins are first dissolved in denaturants such as urea, guanidine–HCl, and acid which keep proteins unfolded and prevent aggregation. It is also important to include reducing reagents such as 2-mercaptoethanol and dithiothreitol (DTT) to keep sulfhydryl groups reduced. When denaturants are removed, proteins are allowed to fold. This is normally accomplished by three methods (80):

(a) Dialysis. Protein concentrations are normally kept high and the denaturant concentration is decreased slowly so that proteins tend to aggregate. Despite this drawback, a number of proteins have been re-folded efficiently by this method (81).

(b) Dilution. Protein in urea or guanidine–HCl solutions are diluted into a large volume of buffer. The denaturant concentration quickly decreases and proteins are allowed to fold. As protein molecules spend little time in intermediate concentrations of denaturant, the aggregation of unfolded protein or folding intermediates can be avoided.

(c) Folding on a solid support. Aggregation of proteins can be avoided by fixing proteins on a solid support. Proteins in urea are applied to an ion-exchange column and as urea is washed from the column proteins are allowed to fold. The proteins are then eluted from the column with a salt gradient (82).

Whichever of the above methods is used, the pH, ionic strength, organic solvent, other additives, temperature, and the time course of changes in these parameters affect the folding process. Optimal conditions must be found for each protein.

Protocol 13. Re-folding of β-globin (17)

Reagents

- β-Globin solution. Dissolve β-globin in 8 M urea, 50 mM Tris–HCl pH 8.0, 1 mM EDTA, 1 mM DTT and dilute with the same buffer to 5 mg/ml.
- α-Globin solution. Dilute a 1.2-fold excess of α-globin (native form) into cold buffer

(50 mM Tris–HCl pH 8.0) of 16 times the volume of the β-globin solution.
- Haemin. Dissolve 25 mg of haemin first in 2.5 ml 1 M KOH to which 20 ml of distilled water and 2.5 ml 1 M KCN are added.

Method

1. Dilute the β-globin solution into the solution containing the folded α-globin with gentle, continuous stirring. By doing this, both the β-globin and urea are diluted 17 times and the β-globin is allowed to fold. The solution is left to stand for 10 min on ice. The unstable β-globin is stabilized by binding to the folded α-globin. The final concentration of urea is less than 0.8 M and the folded protein is stable under these conditions.

2. Add slowly a 1.2-fold excess of haemin-dicyanide over β-globin with gentle stirring using a peristaltic pump over a period of 15 min.

3. Dialyse the solution containing the reconstituted haemoglobin against cold phosphate buffer pH 7.4 to remove the urea completely.

4. Purify the reconstituted haemoglobin by ion-exchange chromatography.

The folding process is more complex for proteins with disulfide bridges. Formation of incorrect disulfide bridges must be avoided. Folding is normally carried out in the presence of mixed disulfide reagents such as a mixture of reduced and oxidized glutathione. Folding of proteins with a prosthetic group such as haem, vitamin B12, and flavin is more difficult to achieve. Such proteins are normally folded first in the absence of the prosthetic group and the prosthetic group is added later to the folded polypeptide. Oligomeric proteins such as haemoglobin are difficult to re-fold. The folding of β-subunits of haemoglobin was accomplished in the presence of the partner α-subunit. The α-subunits stabilize the folded β-globins. As an example, the procedure for re-folding β-globin is given in *Protocol 13*. *In vitro* re-folding of proteins has been discussed more extensively elsewhere and so the reader is referred to this specialized literature (80).

Acknowledgements

We thank Nicola Pavletich for his kind help in using his methods, and our colleagues Jo Avis, Noboru Komiyama, and Marina Parry for helpful discussions and sharing unpublished results.

References

1. Hendrickson, W. A. and Wüthrich, K. (ed.) (1991). *Macromolecular structures 1991*. Current Biology Ltd., London. Hendrickson, W. A. and Wüthrich, K. (ed.) (1992). *Macromolecular structures 1992*. Current Biology Ltd., London.
2. Hartley, D. L. and Kane, J. F. (1988). *Biochem. Soc. Trans.*, **16**, 101.
3. Ikemura, T. (1982). *J. Mol. Biol.*, **158**, 573.
4. Parker, J., Johnston, T. C., Borgia, P. T., Holtz, G., Remaut, E., and Fiers, W. (1982). *J. Mol. Biol.*, **258**, 10007.
5. Springer, B. A. and Sligar, S. G. (1987). *Proc. Natl. Acad. Sci. USA*, **84**, 8961.
6. Hoffman, S. J., Looker, D. L., Roehrich, J. M., Cozart, P. E., Durfee, S. L., Tedesco, J. L., and Stetler, G. L. (1990). *Proc. Natl. Acad. Sci. USA*, **87**, 8521.
7. Lin, K., Kurland, I., Xu, L. Z., Lange, A. J., Pilkis, J., El-Maghrabi, M. R., and Pilkis, S. J. (1990). *Protein Expression and Purification*, **1**, 169.
8. Marston, F. A. O. (1987). In *DNA cloning: a practical approach* (ed. D. M. Glover), Vol. III, pp. 59–88. IRL Press, Oxford.
9. Guide to protein purification. (1990). *Methods in enzymology* (ed. M. P. Deutscher), Vol. 182. Academic Press, London.

10. Komiyama, N. H., Shih, D. T.-B., Looker, D., Tame, J., and Nagai, K. (1991). *Nature*, **352**, 349.
11. Jorin, T. M., Englund, P. T., and Butsch, L. L. (1969). *J. Biol. Chem.*, **244**, 2996.
12. Scopes, R. (1982). *Protein purification: principles and practice*. Springer-Verlag, New York.
13. Grodberg, J. and Dunn, J. J. (1988). *J. Bacteriol.*, **170**, 1245.
14. Nelson, H. C. M. and Sauer, R. T. (1986). *J. Mol. Biol.*, **192**, 27.
15. Nagai, K. and Thøgersen, H. C. (1987). In *Methods in enzymology* (ed. R. Wu and L. Grossman), Vol. 153, pp. 461–81. Academic Press, London.
16. Marston, F. A. O., Lowe, P. A., Doel, M. T., Shoemaker, J. M., Whiter, S., and Angal, S. (1984). *Bio/Technology*, **2**, 800.
17. Jessen, T. H., Komiyama, N. H., Tame, J., Pagnier, J., Shih, D., Luisi, B., Fermi, G., and Nagai, K. (1993). In *Methods in enzymology* (ed. J. Everse, K. D. Vandegrift, and R. M. Winslow), Vol. 231, pp. 347–64. Academic Press, London.
18. Pavletich, N. P. and Pabo, C. (1991). *Science*, **252**, 809.
19. Uhlén, M. and Moks, T. (1990). In *Methods in enzymology* (ed. D. V. Goeddel), Vol. 185, pp. 129–43. Academic Press, London.
20. Ford, C. F., Suominen, I., and Glatz, C. E. (1991). *Protein Expression and Purification*, **2**, 95.
21. Nilsson, B., Forsberg, G., Moks, T., Hartmanis, M., and Uhlén, M. (1992). *Curr. Opinion Struct. Biol.*, **2**, 569.
22. Sassenfeld, H. M. (1990). *Trends Biotechnol.*, **8**, 88.
23. Ullmann, A. (1984). *Gene*, **29**, 27.
24. Scherf, A., Mattei, D., and Schreiber, M. (1990). *J. Immunol. Methods*, **128**, 81.
25. Smith, D. B. and Johnson, K. S. (1988). *Gene*, **67**, 31.
26. Dykes, C. W., Bookless, A. B., Coomber, B. A., Noble, S. A., Humber, D. C., and Hobden, A. N. (1988). *Eur. J. Biochem.*, **174**, 411.
27. Knott, J. A., Sullivan, C. A., and Weston, A. (1988). *Eur. J. Biochem.*, **174**, 405.
28. Anba, J., Baty, D., Lloubes, R., Pages, J.-M., Joseph-Liauzun, E., Shire, D., Roskam, W., and Lazdunski, C. (1987). *Gene*, **53**, 219.
29. Nilsson, B. and Abrahmsén, L. (1990). In *Methods in enzymology* (ed. D. V. Goeddel), Vol. 185, pp. 144–61. Academic Press, London.
30. Nilsson, B., Abrahmsén, L., and Uhlén, M. (1985). *EMBO J.*, **4**, 1075.
31. Moks, T., Abrahmsén, L., Österlöf, B., Josephson, S., Östling, M., Enfors, S.-O., Persson, I., Nilsson, B., and Uhlén, M. (1987). *Bio/Technology*, **5**, 379.
32. Nygren, P.-Å., Eliasson, M., Abrahmsén, L., Uhlén, M., and Palmcrantz, E. (1988). *J. Mol. Recognit.*, **1**, 69.
33. Bedouelle, H. and Duplay, P. (1988). *Eur. J. Biochem.*, **171**, 541.
34. Guan, C., Li, P., Riggs, P. D., and Inouye, H. (1988). *Gene*, **67**, 21.
35. Maina, C. V., Riggs, P. D., Grandea, A. G. III, Slatko, B. E., Moran, L. S., Tagliamonte, J. A., McReynolds, L. A., and Guan, C. (1988). *Gene*, **74**, 365.
36. Blondel, A. and Bedouelle, H. (1990). *Eur. J. Biochem.*, **193**, 325.
37. Krivi, G. G., Bittner, M. L., Rowold, E. Jr., Wong, E. Y., Glenn, K. C., Rose, K. S., and Tiemeier, D. C. (1985). *J. Biol. Chem.*, **260**, 10263.
38. Marks, J. D., Hoogenboom, H. R., Bonnert, T. P., McCafferty, J., Griffiths, A. D., and Winter, G. (1991). *J. Mol. Biol.*, **222**, 581.
39. Hopp, T. P., Prickett, K. S., Price, V. L., Libby, R. T., March, C. J., Cerretti, D. P., Urdal, D. L., and Conlon, P. J. (1988). *Bio/Technology*, **6**, 1204.

40. Prickett, K. S., Amberg, D. C., and Hopp, T. P. (1989). *BioTechniques*, **7**, 580.
41. Kortt, A. A., Caldwell, J. B., Gruen, L. C., and Hudson, P. J. (1992). *IBI FLAG Epitope*, **1**, 9.
42. Sassenfeld, H. M. and Brewer, S. J. (1984). *Bio/Technology*, **2**, 76.
43. Zhao, J., Ford, C. F., Glatz, C. E., Rougvie, M. A., and Gendel, S. M. (1990). *J. Biotechnol.*, **14**, 273.
44. Parker, D. E., Glatz, C. E., Ford, C. F., Gendel, S. M., Suominen, I., and Rougvie, M. A. (1990). *Biotechnol. Bioeng.*, **36**, 467.
45. Dalbøge, H., Dahl, H.-H. M., Pedersen, J., Hansen, J. W., and Christensen, T. (1987). *Bio/Technology*, **5**, 161.
46. Porath, J., Carlsson, J., Olsson, I., and Belfrage, G. (1975). *Nature*, **258**, 598.
47. Hochuli, E., Bannwarth, W., Döbeli, H., Gentz, R., and Stüber, D. (1988). *Bio/Technology*, **6**, 1321.
48. Persson, M., Bergstrand, M. G., Bülow, L., and Mosbach, K. (1988). *Anal. Biochem.*, **172**, 330.
49. Hochuli, E., Döbeli, H., and Schacher, A. (1987). *J. Chromatogr.*, **411**, 177.
50. Arnold, F. H. (1991). *Bio/Technology*, **9**, 151.
51. Skerra, A., Pfitzinger, I., and Plückthun, A. (1991). *Bio/Technology*, **9**, 273.
52. Le Grice, S. F. J. and Güninger-Leitch, F. (1990). *Eur. J. Biochem.*, **187**, 307.
53. Van Dyke, M. W., Sirito, M., and Sawadogo, M. (1992). *Gene*, **111**, 99.
54. Ausubel, F. M., Brent, R., Kingston, R. E., Moore, D. D., Seidman, J. G., Smith, J. A., and Struhl, K. (ed.) (1991). *Current protocols in molecular biology*. John Wiley and Sons, New York.
55. Hoover, D., Friedmann, M., Reeves, R., and Magnuson, N. S. (1991). *J. Biol. Chem.*, **266**, 14018.
56. Ni, L., Guan, K., Zalkin, H., and Dixon, J. E. (1991). *Gene*, **106**, 197.
57. Lew, A. M., Beck, D. J., and Thomas, L. M. (1991). *J. Immunol. Methods*, **136**, 211.
58. Kaelin, W. G. Jr., Pallas, D. C., DeCaprio, J. A., Kaye, F. J., and Livingston, D. M. (1991). *Cell*, **64**, 521.
59. Fikrig, E., Barthold, S. W., Kantor, F. S., and Flavell, R. A. (1990). *Science*, **250**, 553.
60. Wynn, R. M., Davie, J. R., Cox, R. P., and Chuang, D. T. (1992). *J. Biol. Chem.*, **267**, 12400.
61. Sampson, B. A. and Gotschlich, E. C. (1992). *Proc. Natl. Acad. Sci. USA*, **89**, 1164.
62. Louis, J. M., McDonald, R. A., Nashed, N. T., Wondrak, E. M., Jerina, D. M., Oroszlan, S., and Mora, P. T. (1991). *Eur. J. Biochem.*, **199**, 361.
63. Lauritzen, C., Tüchsen, E., Hansen, P. E., and Skovgaard, O. (1991). *Protein Expression and Purification*, **2**, 372.
64. De Vos, P., Claessens, F., Winderickx, J., Van Dijck, P., Celis, L., Peeters, B., Rombauts, W., Heyns, W., and Verhoeven, G. (1991). *J. Biol. Chem.*, **266**, 3439.
65. Grosso, L. E., Parks, W. C., Wu, L., and Mecham, R. P. (1991). *Biochem. J.*, **273**, 517.
66. Hammarberg, B., Nygren, P.-Å., Holmgren, E., Elmblad, A., Tally, M., Hellman, U., Moks, T., and Uhlén, M. (1989). *Proc. Natl. Acad. Sci. USA*, **86**, 4367.
67. Murby, M., Cedergren, L., Nilsson, J., Nygren, P.-Å., Hammarberg, B., Nilsson, B., Enfors, S.-O., and Uhlén, M. (1991). *Biotechnol. App. Biochem.*, **14**, 336.

68. Hanada, K., Yamato, I., and Anraku, Y. (1988). *J. Biol. Chem.*, **263**, 7181.
69. Munro, S. and Pelham, H. R. B. (1986). *Cell*, **46**, 291.
70. Evan, G. I., Lewis, G. K., Ramsay, G., and Bishop, J. M. (1985). *Mol. Cell. Biol.*, **5**, 3610.
71. Gross, E. (1967). In *Methods in enzymology* (ed. C. H. W. Hirs), Vol. 11, pp. 238–55. Academic Press, London.
72. Stark, G. R. (1977). In *Methods in enzymology* (ed. C. H. W. Hirs and S. N. Timasheff), Vol. 47, pp. 129–32. Academic Press, London.
73. Landon, M. (1977). In *Methods in enzymology* (ed. C. H. W. Hirs and S. N. Timasheff), Vol. 47, pp. 145–9. Academic Press, London.
74. Bornstein, P. and Balian, G. (1977). In *Methods in enzymology* (ed. C. H. W. Hirs and S. N. Timasheff), Vol. 47, pp. 132–45. Academic Press, London.
75. Smyth, D. G. (1967). In *Methods in enzymology* (ed. C. H. W. Hirs), Vol. 11, pp. 214–31. Academic Press, London.
76. Mitchell, W. M. (1977). In *Methods in enzymology* (ed. C. H. W. Hirs and S. N. Timasheff), Vol. 47, pp. 165–70. Academic Press, London.
77. Belagaje, R. M., Mayne, N. G., Van Frank, R. M., and Rutter, W. J. (1984). *DNA*, **3**, 120.
78. Carter, P., Abrahmsén, L., and Wells, J. A. (1991). *Biochemistry*, **30**, 6142.
79. Varadarajan, R., Szabo, A., and Boxer, S. G. (1989). *Proc. Natl. Acad. Sci. USA*, **82**, 5681.
80. Jaenicke, R. and Rudolph, R. (1989). In *Protein structure: a practical approach* (ed. T. E. Creighton), pp. 191–223. IRL Press, Oxford.
81. Reinach, F. C., Nagai, K., and Kendrick-Jones, J. (1986). *Nature*, **322**, 80.
82. Creighton, T. E. (1986). Patent WO 86/05809.
83. Laemmli, U. K. (1970). *Nature* (London), **227**, 680.
84. Cronan, J. E. Jr. (1990). *J. Biol. Chem.*, **265**, 10327.
85. Cress, D., Shultz, J., and Breitlow, S. (1993) *Promega Notes*, **42**, 2.
86. Consler, T. G., Persson, B. L., Jung, H., Zen, K. H., Jung, K., Privé, G. G., Verner, G. E., and Kaback, H. R. (1993). *Proc. Natl. Acad. Sci. USA*, **90**, 6934.
87. Germino, F. J., Wang, Z. X., and Weissman, S. M. (1993). *Proc. Natl. Acad. Sci. USA*, **90**, 933.
88. Schmidt, T. G. M. and Skerra, A. (1993). *Protein Eng.*, **6**, 109.

Production of monoclonal antibodies against proteins expressed in *E. coli*

STEVEN M. PICKSLEY and DAVID P. LANE

1. Introduction

The production of monoclonal antibodies against the product of a genetically identified open reading frame (ORF) is often a major starting point in its characterization. The procedure is more complex than that described in Chapter 2 for the production of polyclonal antibodies and should be seen as a complementary technique. Ideally, both procedures should be employed to exploit the full potential of immunochemical analysis. For while the major benefits of monoclonal antibodies, namely their unlimited supply and defined specificity, come into their own during the advanced stages of characterization, polyclonal antibodies do allow simple detection and immunodepletion studies. The preparation of monoclonal antibodies is, however, preferred for the quantitative assay of the ORF product and for rapid purification of the product in a biologically active form. Moreover, the fine mapping of the antibody binding sites (epitopes) on the ORF product, when correlated with their effect on biological activity, provides a valuable insight into possible structure–function relationships.

In order to raise monoclonal antibodies, sufficient quantities of the immunogen are required to elicit a strong immune response. Since it is not always possible to prepare large amounts of native protein, the ORF is cloned into a suitable *E. coli* expression vector to allow over-expression of the product and facilitate its purification. This approach also allows smaller defined regions of the ORF to be expressed for the production of antibodies against specific regions of the ORF, and the subsequent fine mapping of epitopes. A summary of the the steps involved in the production and uses of monoclonal and polyclonal antibodies raised against ORFs expressed in *E. coli* is shown in *Figure 1*.

There is a wide variety of bacterial expression vectors available (see Chapter 2). Generally, they are based on plasmid vectors with strong

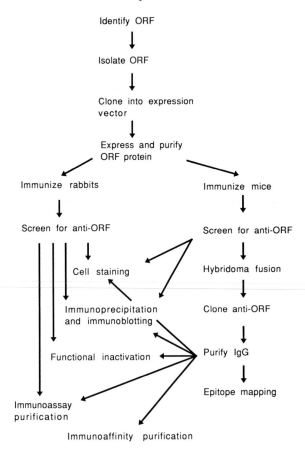

Figure 1. Diagram showing the scheme for the production and uses of monoclonal and polyclonal antibodies raised against ORFs expressed in *E. coli*.

promoters and antibiotic selection systems. One such plasmid used extensively in our laboratory is the pT7-7 plasmid, which has a bacteriophage T7 promoter (from the T7 ϕ*10* gene), and encodes resistance to ampicillin (1). The ORF is cloned downstream of the T7 ϕ*10* gene start site, see *Figure 2*. The construct is then transfected into a host strain in which expression of the T7 RNA polymerase is regulated by a *lacI* encoded repressor (2), such that expression of the T7 RNA polymerase, and concomitantly that of the ORF, can be induced by isopropyl-β-D-thiogalactopyranoside (IPTG). Currently, there are similar vectors to the pT7-7 plasmid which have been engineered to fuse small tags to the ORF product so that they can be recognized by commercially available monoclonal antibodies or allow one-step purification on metal ion columns (see Chapters 2 and 3). Both of these developments facilitate the identification and purification of the ORF product, particularly in cases where

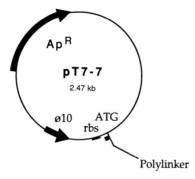

Figure 2. Map of the pT7-7 vector; Ap^R indicates the β-lactamase gene which confers ampicillin resistance, φ10 is the T7 gene promoter, rbs is a ribosome binding site.

expression is poor. Another advantage of using these expression systems is that the construct can easily be modified by appropriate restriction digests or polymerase chain reaction (PCR) amplification to just express regions that are known to be immunologically silent in the native protein or of defined functional interest.

2. Cloning and expression of gene fragments

Most of the methods and reagents described in this section are based on protocols found elsewhere (3).

2.1 Construction of the ORF expression plasmid

The method by which the ORF was identified will determine the initial approach. Where the starting point is a just a nucleotide sequence, the initial approach will involve amplifying DNA from cDNA using PCR. The PCR primers should be designed to include a restriction endonuclease site to allow the ORF to be cloned into a site in-frame with the initiating ATG of the construct. Care should be taken to ensure that the restriction sites in the primers are in the middle of the primers such that these do not interfere with priming DNA synthesis from the 3' end of the oligonucleotide primers, and are readily cleavable by the restriction enzymes (the efficiency of digestion by various enzymes for 5' and 3' ends of stated length is published in the New England Biolabs Catalogue). Should problems be experienced in trying to clone the PCR amplified product following digestion with restriction enzymes, try to subclone the product directly into a different vector with colour identification of positive clones, for example using pUC18/19, pBluescript II (Stratagene), or pTA cloning system (Invitrogen), and then subclone into your vector.

For illustration, in *Figure 3*, we present the results of cloning the human tumour suppressor p53 gene into the pT7-7 vector. The figure shows the

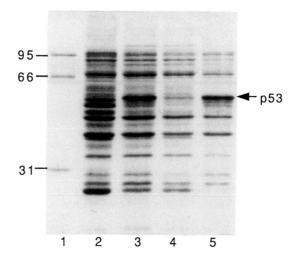

95 —
66 —

← p53

31 —

1   2   3   4   5

Figure 3. IPTG induction of expression of human p53 from *E. coli* cells containing a pT7-7/T7 φ10-human p53 construct. Cell lysates were fractionated by 10% SDS–PAGE and visualized by staining with Coomassie blue R250. Lane 1, protein molecular weight markers ($M_r \times 10^3$); lane 2, uninduced cells; lane 3, IPTG induced cells; lane 4, soluble protein fraction; lane 5, insoluble fraction. The samples were prepared essentially as described in *Protocol 2*. The position of p53 is indicated. The figure is taken from ref. 4, with permission.

IPTG induced expression of the human p53 gene from the T7 φ*10* promoter of pT7-7. In this example (4), the over-expressed p53 protein is visible in the total cell protein (lane 3), and following cell lysis is found in the insoluble cell pellet (lane 5). *Protocols 1* and *2* describe how these samples were generated. The insoluble p53 protein was subsequently solubilized in 5 M guanidine–hydrochloride and used to raise the polyclonal sera CM1 and the monoclonal antibodies DO-1 and DO-7 (4, 5). All of these reagents have been widely used in the immunochemical analysis of p53 in both molecular and clinical contexts (see *Figure 4*).

Protocol 1. Preparation of samples for SDS–polyacrylamide gel electrophoresis [a]

Reagents

- L broth containing 50 μg/ml ampicillin
- 100 mM IPTG in water
- 2 × sample buffer stock: 125 mM Tris–HCl

pH 6.8, 2% (w/v) SDS, 10% glycerol, 10% β-mercaptoethanol or 100 mM DTT, 0.001% (w/v) bromophenol blue

Method

1. Grow up a 2 ml overnight culture in L broth containing 50 μg/ml ampicillin.

2. Inoculate 5 ml of L broth containing 50 µg/ml of ampicillin with 0.1 ml of overnight culture and grow to early to mid-log phase (OD_{600} of 0.2–0.6), and then add IPTG to 0.5 mM for 1–2 h with aeration.

3. Transfer 1.5 ml of culture to a 1.5 ml microcentrifuge tube.

4. Spin for 1 min in a microcentrifuge (15 000 *g*) and aspirate off the supernatant.

5. Resuspend the pellet in 200 µl of 1 × sample buffer.

6. Boil for 5 min before loading 5–25 µl of the sample on the gel.

[a] This protocol is for strain BL21 (DE3) and pT7-7 constructs, and should be modified for other vectors; for example, the antibiotic supplement to the L broth will vary with the vector used. For any new expression construct it is important to establish the optimum induction conditions by varying the period of induction and the concentration of the inducing agent.

Protocol 2. Determination of whether the ORF protein is expressed in a soluble or insoluble form

Equipment and reagents

- Sonicator (optional)
- Lysis buffer: 50 mM Tris–HCl pH 8.0, 2.0 mM EDTA, 100 mM NaCl, containing 0.25 mg/ml lysozyme
- 10% Triton X-100

- 1 M $MgCl_2$
- 2 × sample buffer (*Protocol 1*)
- 1 mg/ml pancreatic DNase I (or DNase of choice)

Method

1. Induce expression of the ORF protein in a 5 ml culture as described in *Protocol 1*.

2. Harvest the cells by centrifugation at 4000 *g* for 5 min at 4 °C.

3. Resuspend the cell pellet in 1 ml of 50 mM Tris–HCl pH 8.0, 2.0 mM EDTA, 100 mM NaCl, containing 0.25 mg/ml lysozyme. Incubate at 30 °C for 15 min and then add Triton X-100 to 0.2% (v/v).

4. Degrade the DNA either by sonication for two 10 sec bursts, or add $MgCl_2$ to 10 mM and pancreatic DNase I to 1 µg/ml. Incubate for 30 min at room temperature.

5. Centrifuge at 4000 *g* for 5 min at 4 °C.

6. Resuspend the pellet in 250 µl of 1 × sample buffer, and add 250 µl of 2 × sample buffer to 250 µl of supernatant.

7. Boil for 5 min before loading 5–25 µl of the supernatant and pellet samples on the gel.

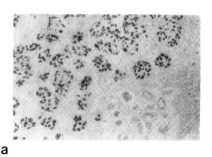
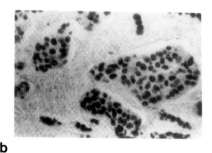

a   b

Figure 4. Immunohistochemical detection of the tumour suppressor protein p53 by monoclonal antibody DO1. Immunohistochemical detection of p53 by antibody DO1 on frozen (a)
and paraffin sections (b) of human breast carcinoma. The stained nuclei identify cancer cells
that accumulate an abnormal p53 protein, since p53 protein is not usually detectable in
normal cells by immunohistochemistry. The figure is taken from ref. 5, with permission.

2.2 Selection of vector

In the previous example the vector chosen for the expression of p53 was
pT7-7. Since that construct was made, a wider number of other vectors have
become available that not only facilitate the over-expression of ORF proteins
in *E. coli* but also assist in their purification. These vectors and their sources
are mentioned below but the reader is referred to Chapter 2 for detailed
descriptions of the pGEX, pET, and pMAL vectors and their uses.

(a) His-tag vectors, pET series (Novagen/AMS Biotechnology), and pQE
series (Hybaid Ltd). These vectors are quite diverse, but are are all based
on placing repeats of histidine residues at the N- or C-terminal regions of
the ORF protein to allow the protein to be purified on the basis that the
ORF protein will now bind to divalent ions (Ni^{2+}) immobilized on a resin
column and can be eluted with imidazole buffers (6). This purification
step works for native proteins and also proteins solubilized in 6 M
guanidinium–hydrochloride.

(b) pGEX vectors (Pharmacia). The ORF is cloned into one of two vectors,
pGEX-2T and pGEX-3X, in-frame with the C-terminal end of a glutathione S-transferase gene under the regulation of a P_{tac} promoter (7).
The resulting fusion protein is purified by affinity chromatography using
glutathione–Sepharose 4B. The ORF protein can then be released from
the fusion protein by cleaving with thrombin or factor Xa.

(c) pMAL vectors (New England Biolabs). In this system the ORF is cloned
downstream of a gene encoding a maltose binding protein and is expressed
from the P_{tac} promoter. The fusion protein is purified by its ability to bind
an amylose column and eluted off with maltose (8). The ORF can be
isolated from the fusion protein by cleaving the fusion protein with
protease factor Xa.

(d) pRIT vectors (Pharmacia). The ORF is cloned into the C-terminal end of a truncated protein A gene that encodes the IgG binding domains enabling the fusion protein to be purified by its ability to bind to IgG–Sepharose 6FF (9). There are two pRIT vectors, pRIT2T and pRIT5. The pRIT2T vector was designed to allow temperature inducible expression of intracellular fusion proteins in *E. coli*, from a lambda P_R promoter in an *E. coli* host strain containing a temperature-sensitive lambda *cI*857 repressor. The other vector, pRIT5, was designed to translocate fusion proteins to the periplasm in *E. coli* and to be secreted across the cell membrane in *Staphylococcus aureus*.

3. Production of monoclonal antibodies

Cell lines producing monoclonal antibodies are usually derived from mice or rats that show a good immune response to an immunogen. A suspension of their spleen cells is prepared and the cells fused with a myeloma cell line. The fused cells produced thus exhibit characteristics derived from both parental cells, combining the property of unlimited cell division of the myeloma parent with the property of specific antibody production of the spleen B cell. The cell fusions producing the monoclonal antibody of potential interest (hybridomas) are then cloned by limiting dilution to remove contaminating hybridomas, and also to select for genetically stable cell lines.

The cloning of ORFs in protein expression vectors permits immunization with large amounts of pure protein, and this makes the production of monoclonal antibodies directed to sequences within that ORF a straightforward, if somewhat laborious, task. The primary practical requirements are for good tissue culture facilities and a sensitive screening assay that will distinguish antibodies directed against the ORF from those that recognize any vector derived sections of the protein. It is the establishment of this assay that is the chief stumbling block in the procedure. Once ORF-specific monoclonal antibodies have been obtained, they can be used to identify, quantitate, and purify the product of the ORF.

3.1 Preparation of the antigen

The purification of ORF proteins has been described in Chapters 2 and 3 and in *Protocol 2*. The amounts of ORF protein needed for immunization of mice are relatively small (50–250 μg in total) so all methods of preparation are suitable. In *Protocol 3* a method is described for isolating protein from SDS–polyacrylamide gels, and is the method of choice where the ORF protein is abundant. When the ORF protein is not abundant, the excised region of the gel may be used directly by emulsifying with complete Freund's adjuvant to immunize the mouse. A protocol is given in Chapter 2 (*Protocol 10*). Whilst material derived from preparative SDS–polyacrylamide gels may not be as immunogenic as a more native preparation, it does have the advantage of inducing a response to denaturation-resistant epitopes. Monoclonal antibodies directed to such epitopes are often disproportionately useful in the subsequent detailed immunochemical analysis of protein antigens. Absolute purity of the immunogen is not required as only monoclonal antibodies directed to epitopes specific to the cloned insert will be selected.

Protocol 3. Purification of the ORF protein by preparative SDS–polyacrylamide gel electrophoresis

Equipment and reagents

- Vertical SDS–PAGE gel electrophoresis equipment
- Horizontal gel electrophoresis equipment
- SDS–PAGE gel (the acrylamide concentration depends on the size of the ORF protein to be fractionated)
- Pre-stained protein molecular weight markers (optional)

- SDS–PAGE gel electrophoresis running buffer: 0.025 M Tris-HCl pH 6.8, 0.19 M glycine, 0.1% SDS
- 0.2% (w/v) Coomassie blue R250 stain in 50% (v/v) methanol, 10% (v/v) acetic acid
- Destain solution 1: 50% (v/v) methanol, 10% (v/v) acetic acid
- Destain solution 2: 7.5% (v/v) methanol, 10% (v/v) acetic acid

Method

1. Run the sample on an SDS–polyacrylamide gel in a single large slot adjacent to a lane with pre-stained protein molecular weight markers.

2. Remove a vertical strip of the gel containing the ORF protein and place the strip in destain solution 1 for 5 min at room temperature. Ensure the remainder of the gel does not dry out by placing it in SDS–PAGE running buffer.

3. Stain the strip with 0.2% Coomassie blue R250 stain for 5 min at room temperature.

4. Destain the strip with destain solution 2 at room temperature for as long as required to visualize the ORF protein.

5. Line up the remaining gel, and excise the region where the ORF protein has migrated. Place the gel slice in a dialysis bag containing SDS–PAGE running buffer.

6. Using a horizontal gel electrophoresis apparatus and SDS–PAGE gel electrophoresis running buffer, electroelute the protein out of the gel.

7. Store the protein in aliquots at −20 °C.

8. Determine the recovery and concentration of the ORF protein by comparison with known standards by SDS–PAGE.

3.1.1 Immunization protocol

The aim of the initial immunizations is to obtain a detectable antibody response to the ORF sequence and to use the resulting polyclonal sera to

Table 1. Immunization schedule

Day[a]	Injection	Adjuvant[b]	Test bleed
1	—	—	Pre-bleed
3	50 μg	Complete Freund's adjuvant	—
24	50 μg	Complete Freund's adjuvant	—
34	—	—	Test-bleed
35	50 μg	Incomplete Freund's adjuvant	—
42	—	—	Test-bleed
43	50 μg	Incomplete Freund's adjuvant	—
50	50 μg	Incomplete Freund's adjuvant	—
57	—	—	Test-bleed

[a] This protocol may be continued as required.
[b] The volume of all injections is 200 μl, consisting of 100 μl of protein in PBS (500 μg/ml) and 100 μl of adjuvant mixed by passing through a double Luer fitting using two syringes. The injection is given intraperitoneally, and is subject to the national or local legislation concerning the use of animals in scientific procedures.

establish an ORF-specific screening assay. For routine use, three to six month-old female BALB/c mice are suitable. The initial two injections should be given in Freund's complete adjuvant, and subsequent injections in incomplete adjuvant, except for the final pre-fusion boost which should be given in the absence of adjuvant (see *Table 1*). All injections can be given intraperitoneally. Mice should be immunized in groups of three to six animals to give a reserve of fusion donors should a fusion fail. A suitable immuniza-tion schedule is given in *Table 1*. Whilst a response is usually detected after two injections, antibodies specific to the ORF may not appear until much later. It is therefore worth persevering with an immunization schedule for at least seven injections before concluding that the immunogen is unsuitable.

3.1.2 Detection of ORF-specific antibodies

It is vital to establish that antibodies specific to ORF sequences are present in the sera of the immunized mice. There are a number of distinct approaches that can be taken, depending on the availability of the full-length native product of the ORF. If it is known that the ORF is expressed by a tissue culture cell line or is present at reasonable levels in a particular tissue, then the reactivity of the sera with the cell line or tissue in immunocytochemical, immunoprecipitation, or Western blotting procedures can be measured. In our experience with mammalian and viral ORFs, the crude polyclonal sera can be used in all three of these methods without serious background prob-lems. Control sera should include the pre-bleed of the test mouse and hyper-immune sera directed to the remainder of the fusion protein where possible. If nothing is known about the expression of the ORF, then an alternative strategy must be adopted that exploits the available ORF sequence. One approach for ORF fusion proteins is to remove antibodies to the non-ORF part of the fusion protein selectively by immunoabsorption and then to test the unbound fraction in a Western blot procedure for reaction with the fusion protein but not with the non-ORF part of the fusion. An alternative approach is to subclone the ORF into an alternative vector that does not share any homology in the non-ORF part of the fusion. The reactivity of the sera with the new construct either in colony blot or Western blot procedures will then confirm the presence of antibodies specific to the ORF. Again, tests with the same control sera are required, and in addition, all sera should be checked for reactivity with the new vector alone. The convincing establishment of an anti-ORF specific response by at least one of these methods is an absolute pre-requisite before proceeding to the next stage of preparing hybridoma cells.

3.2 Setting-up the hybridoma screen

Successful fusions demand rapid, efficient screens. Most fusions will need to be screened three times, and since each fusion is plated out over 500–1000 wells, this means that up to 3000 separate samples need to be tested for specific antibody production. Furthermore, the kinetics of the growth of the hybridoma cells demand that the results of the screening test be known within 48 hours of sampling. The only practical way to handle this kind of sample is to use an assay based in microtitre plates that gives a direct visual readout. The screening assay need not be absolutely definitive and some false positives can be tolerated. The assay can then be backed up by a more definitive test which need only be carried out on the relatively small number of samples that scored positive in the initial screening assay. One of the most useful types of screening assay is an ELISA test, in which antigen is coated on to the bottom of plastic microtitre wells and bound antibody present in the hybridoma sample is detected with an enzyme-conjugated anti-immunoglobulin. A basic protocol for this assay is presented in *Protocol 4*.

Protocol 4. ELISA protocol for protein antigen

Equipment and reagents

- Plastic microtitre plates (Falcon 3912, Becton Dickinson)
- Multichannel dispensers (an eight channel 5–50 µl, and an eight channel 50–300 µl)
- ELISA plate reader (not absolutely essential)
- Phosphate-buffered saline solution (PBS). Dissolve 8.0 g of NaCl, 0.2 g of KCl, 1.44 g of Na_2HPO_4, and 0.24 g of KH_2PO_4 in 800 ml of distilled water. Adjust the pH to 7.2 and then add water to make one litre of solution. Autoclave, and then store at room temperature.
- Protein antigen solution: 0.1–1.0 µM in PBS
- Blocking solution: 3% BSA in PBS
- Wash solution: 0.1% NP-40 in PBS

- Secondary antibody. This is rabbit antibodies to mouse immunoglobulins coupled to horse-radish peroxidase (DAKO), diluted 1:1000 in PBS containing 10% fetal calf serum (Advanced Protein Products).
- Substrate solution. Dissolve 3,3',5,5' tetramethyl benzidine (T-2885, Sigma) in dimethyl sulfoxide at a concentration of 1 mg/ml. Dilute 1/100 in PBS and filter through Whatman No. 1 paper. Finally add hydrogen peroxide to a concentration of 0.01%.
- 70% ethanol
- Stop solution: 1 M H_2SO_4

Method

1. Add 50 μl of protein antigen solution to each well of a microtitre plate and incubate overnight at 4 °C.

2. Remove the protein solution. The plates can be now stored at −20 °C almost indefinitely.

3. Block the plates before use by incubating with 200 μl of blocking solution per well for 1 h at room temperature.

4. Remove the blocking solution, rinse the plates in PBS, and add hybridoma supernatant samples at 2–100 μl per well by direct transfer from the fusion master plates using the multichannel pipette. Rinse the tips between rows in 70% ethanol, and change the tips between plates. Incubate for 3 h at room temperature.

5. Wash three times with 200 μl of wash solution and add the enzyme-conjugated second antibody solution at 50 μl per well. Incubate for 1 h, then wash three times with wash solution and once in PBS.

6. Add 50 μl of freshly prepared substrate solution per well and incubate for a suitable period. Stop the chromogenic reaction by the addition of 50 μl of stop solution per well.

7. Read the plates in the ELISA plate reader. The results are most conveniently analysed by a matrix plot in which the values for optical density are ranked on a scale of 1–10.

An alternative screening test which we have found to be surprisingly versatile is the direct immunocytochemical staining of fixed cells. This is described in *Protocol 5*, and an example is shown in *Figure 4*. The advantage of this method is that only wells that are visibly darkened on the addition of substrate need to be examined under the microscope, and the localization of the substrate deposition can give a great deal of information about the specificity of the antibody. Since the plates can be stored at −20 °C after fixation, it is quite easy to screen a given fusion against a number of different cell lines. The range of the assay can be extended by using virus infected or stably transfected cell lines.

Protocol 5. Immunocytochemical staining of cells

Equipment and reagents

- Microscope
- Tissue culture dishes (90 mm diameter) for screening mouse sera, or microtitre plates for screening hybridomas
- PBS (*Protocol 4*)
- Acetone/methanol (1:1)
- Rabbit anti-mouse immunoglobulin coupled to horse-radish peroxidase, diluted 1:100 in PBS containing 1% bovine serum albumin, 10% fetal calf serum

- 3% (w/v) nickel sulfate
- 30% H_2O_2
- DAB stain solution. Dissolve 10 mg of 3,3'-diaminobenzidine-tetrahydrochloride in 20 ml of PBS, add nickel sulfate to 0.03% (w/v). Filter this solution before adding 4 μl 30% H_2O_2
- 0.2% (w/v) sodium azide

Method

1. For the screening of mouse sera, plate the cells out on 90 mm tissue culture dishes marked on their backs with a 10 × 10 grid drawn with a marker pen. For the screening of hybridomas, plate the cells out in the microtitre wells at approximately 10^4 cells/well in a volume of 100 μl. Leave to settle overnight at 37°C.

2. Rinse the cells in PBS and fix them by incubating in acetone/methanol (1:1) for 2 min. Rinse in the same fixative and allow to dry.

3. Incubate the fixed cells with the crude monoclonal antibody supernatant or dilutions of mouse sera for 3 h in a humid environment (for the microtitre plate, use 10–50 μl of supernatant or diluted sera per well; for the 90 mm dish, spot 2 μl of each antibody on to the monolayer over a marked square on the grid).

4. Rinse five times with PBS. Add the diluted rabbit anti-mouse immunoglobulin coupled to horse-radish peroxidase and incubate for 1–3 h in a humid environment.

5. Rinse five times with PBS and add freshly filtered DAB stain solution to which has been added hydrogen peroxide (see *Equipment and reagents*). Leave for 10–30 min in the dark.

6. Wash the cells with water. Pour off the water, and finally overlay the dish with 0.2% sodium azide solution for storage.

Once the initial screen has been established, it is worthwhile devising some supplementary screens that will definitely identify the antibody as being of the required specificity. These screens do not have the same restrictions placed on them as the preliminary screen and can utilize more laborious methods suitable for only small sample numbers. The Western blot and immunoprecipitation protocols (*Protocols 6* and *7*) are both suitable and ideally each should be used against two different expression constructs.

Protocol 6. Western blotting

Equipment and reagents

- Electroblot apparatus
- Pre-stained protein molecular weight markers (optional)
- Nitrocellulose sheet BA85, 0.45 μm (Schleicher and Schuell)
- PBS (*Protocol 4*)
- Screening antibody[a]

- Blocking buffer: 3% BSA in PBS
- Pre-absorbed secondary antibody[b]
- PBS containing 1% (v/v) NP-40
- 50 mg/ml 4-chloro-1-naphthol in ethanol
- 30% (v/v) H_2O_2

Method

1. Fractionate the polypeptides in extracts by electrophoresis on an SDS–polyacrylamide gel. Include pre-stained protein molecular weight standards, if necessary.

2. Transfer the proteins from the gel to a nitrocellulose sheet using an electroblot apparatus, as directed by the manufacturer.

3. Incubate the nitrocellulose sheet in blocking buffer for 1 h.

4. Rinse the nitrocellulose sheet in PBS and cut it into appropriate strips if required. Incubate each sheet in the screening antibody for 3–18 h.

5. Rinse the nitrocellulose in PBS, followed by two 15 min washes in PBS containing 1% NP-40, and two 15 min washes in PBS.

6. Incubate the nitrocellulose in pre-adsorbed second antibody coupled to horse-radish peroxidase for 3–18 h.

7. Rinse the nitrocellulose in PBS, followed by two 15 min washes in PBS containing 1% NP-40, follwed by two 15 min washes in PBS.

8. Freshly dilute the 50 mg/ml 4-chloro-1-naphthol solution 1:100 in PBS and filter it. Add 30% (v/v) hydrogen peroxide at a dilution of 1:5000. Use this final solution to stain the filter.

9. Once staining has occurred sufficiently, rinse the filter in PBS and dry. Keep in the dark.

[a] The screening antibody could be a monoclonal antibody or polyclonal sera.
[b] It is advisable to pre-adsorb the polyclonal antibody with a lysed *E. coli* extract (*Protocol 2*) to reduce non-specific staining. The second antibody should be specific for the species used as host in the preparation of the screening antibody, e.g. rabbit anti-mouse immunoglobulin coupled to horse-radish peroxidase (DAKO) and goat anti-rabbit immunoglobulins, for monoclonal and rabbit polyclonal antibodies, respectively.

Protocol 7. Immunoprecipitation[a]

Equipment and reagents

- Rotating motorized wheel (blood cell mixer): for mixing the contents of microcentrifuge tubes
- Tris wash buffer; 137 mM NaCl, 5 mM KCl, 0.7 mM Na$_2$HPO$_4$, 25 mM Tris-HCl pH 7.4
- Tris wash buffer containing 10 mM EDTA
- NET buffer: 150 mM NaCl, 5 mM EDTA pH 8.0, 50 mM Tris–HCl pH 8.0, 1% NP-40
- NET buffer containing 0.5 M NaCl (composition as for NET buffer but with increased NaCl)
- Specific antibody (monoclonal antibody supernatant, mouse serum, ascites fluid, or purified antibody)
- Sepharose CL4b bead slurry[b] (Pharmacia). Wash the Sepharose beads in NET buffer and resuspend at 10% (v/v) in NET buffer.
- Protein A- or protein G-coupled Sepharose CL4b bead slurry[b] (Pharmacia). Wash the Sepharose beads in NET buffer and resuspend at 10% (v/v) in NET buffer.

A. *Preparation of cell extract*

1. Rinse the cells in Tris wash buffer. Then add more Tris wash buffer containing 10 mM EDTA, and incubate for 1–15 min at 37 °C.

2. Collect the cells by centrifugation at 1000 *g* for 5 min and lyse by incubating in NET buffer for 30 min at 4 °C, with occasional gentle mixing.

3. Centrifuge at 15 000 *g* at 4 °C for 30 min, or in a microcentrifuge for 5 min at 4 °C to remove cell debris.

4. Pre-clear the supernatant of non-specifically reacting material by incubating the preparation with 100 μl of protein A-coupled Sepharose CL4b bead slurry[b] and 10 μl of Sepharose CL4b bead slurry for 30 min on ice with occasional mixing.

5. Centrifuge for 30 sec in a microcentrifuge at 10 000 *g* to remove non-specifically reacting material and transfer the supernatant to a fresh tube for use in the immunoprecipitation assay.

B. *Immunoprecipitation*

1. Perform each immunoprecipitation reaction with a volume of cell supernatant equivalent to approximately 3 × 10^6 cells. In a microcentrifuge tube, incubate the cell extract with 10–100 μl of monoclonal antibody supernatant, or 0.5–5 μl of mouse serum, or 0.1–1 μl of ascites fluid, or 5 μg of pure antibody at 4 °C for 45–60 min.

2. Add 100 μl of protein A- or protein G-coupled Sepharose bead slurry[b] and mix on a rotating wheel for 15 min.

3. Pellet the immune complexes bound to the beads by centrifugation for 30 sec in a microcentrifuge at 10 000 *g* and remove the supernatant.

4. Wash the beads four times with 0.6 ml of NET buffer containing 0.5 M NaCl.

5. Resuspend the pellet in 50 μl of 1 × SDS–polyacrylamide gel sample buffer (*Protocol 1*), boil for 3 min, and centrifuge for 30 sec in a microcentrifuge at 10 000 *g*. Transfer the supernatant to a fresh tube and boil for 3 min.

6. Resolve the proteins on an SDS–polyacrylamide gel and carry out a Western blot (*Protocol 6*).

[a] This protocol should be adapted as required for the protein of interest and should be carried out in parallel with appropriate irrelevant antibody controls.
[b] The choice of protein A or G depends on whether the antibody is subclass IgG1 or not. Where the antibody subclass is not known, use protein G coupled to Sepharose bead slurry.

3.3 Fusion strategy

Once a clear-cut anti-ORF specific response has been established in the immunized mice, preparation for the fusions can begin. Whilst a large number of empirical observations have been recorded about crucial factors that affect the success of fusions, there is still a background of variability in the technique that makes it worthwhile doing several fusions. It is the cloning and isolation of the antibodies and their characterization that requires most effort so working only with fusions that have a large number of healthy growing colonies (more than a 100) is advisable.

3.4 The final boost

Prior to fusion the donor mouse should be given a final boost of antigen. The object of this booster injection is to stimulate selectively the proliferation of antigen reactive B cells and to localize them to the donor organ, in this case the spleen. The boost should not be given when there are still high concentrations of serum antibodies to the immunogen present. The immunized animal should therefore be allowed to rest for at least a month before the boost. If the ORF protein is soluble, the final boost can be given by intravenous injection into a tail vein as this is particularly effective. Inject about 10–100 μg of protein in a volume of 0.5 ml of saline or PBS without adjuvant. Ensure the sterility of the antigen by filtering through a 0.22 μm filter before injection. If the ORF protein is poorly soluble, do not filter or attempt an intravenous injection, but instead inject the unfiltered material intraperotineally six, five, and four days prior to fusion. Fusions should then take place 72–96 hours following the final boost.

3.5 Preparation of the myeloma cells

The hybridoma parent cells must be in very good condition for the fusion and should therefore be obtained from a cell culture reference centre (e.g. American National Type Culture Collection) or from a successful hybridoma laboratory. It is best not to keep them in culture for prolonged periods but to work from a frozen stock kept in liquid nitrogen. The cells should be thawed at least ten days before the fusion so that they have been subcultured at least twice before use in Dulbecco's modified Eagle's medium (DMEM) containing 10% fetal calf serum (FCS). Dilute the cells one in three into fresh medium (DMEM containing 20% FCS) the evening before fusion. Prepare the cells for fusion by washing twice in DMEM without serum and finally resuspend the cells in 5 ml DMEM. You will need 10^7 cells per fusion ($\sim 2 \times 90$ mm plates of subconfluent cells). Both SP2/0-AG14 and NS-1 myeloma lines work well.

3.6 Preparation of the donor spleen cells for the fusion

This procedure is described in *Protocol 8*.

Protocol 8. Preparation of the donor spleen cells

Equipment and reagents

- Low speed centrifuge
- Laminar air flow hood
- Sterile dissection instruments
- Sterile 90 mm Petri dish
- 19 Gauge needles and 5 ml syringes
- Serum-free DMEM (Dulbecco's modified Eagle's medium)
- Mouse immunized with ORF protein

Method

1. Kill the mouse by cervical dislocation.

2. Soak the fur in ethanol and, using sterile instruments and working in a laminar air flow hood, remove the spleen to a sterile 90 mm Petri dish containing 2 ml of serum-free DMEM.

3. Using a pair of 19 gauge needles mounted on 5 ml syringes, tease the spleen apart very thoroughly.

4. Add 15 ml of serum-free DMEM and transfer the cells to a centrifuge tube. Let the cell suspension settle for about 5 min, then carefully decant the suspension free of any particulate matter to a fresh tube.

5. Wash the cells twice in serum-free DMEM by centrifugation at 400 *g* for 5 min at room temperature.

6. Resuspend the spleen cells in 5 ml of serum-free DMEM.

3.7 The fusion

The procedure for fusing donor spleen cells and myeloma cells is described in *Protocol 9*.

Protocol 9. Production of hybridomas

Equipment and reagents

- Low speed centrifuge
- Laminar air flow hood
- CO_2 incubator at 37 °C
- Sterile dissecting instruments
- Inverted microscope
- Multichannel pipette and sterile tips
- Tissue culture plasticware, pipettes, 96-well microtitre dishes, and plastic round-bottom centrifuge tubes
- Fetal calf serum (FCS); preferably purchased from a source that screens batches of serum for their ability to support hybridoma growth, e.g. Advanced Protein Products
- Polyethyene glycol 1500 solution (a pre-screened 50% (w/v) solution in 75 mM Hepes, is available from Boehringer-Mannheim GmbH, No. 783641)

- Selection medium: either HA (hypo-xanthine/azaserine) or HAT (hypoxanthine/aminopterin/thymidine). We prefer HA as it allows more vigorous initial growth of the fused cells. Prepare it as a 100 × stock by dissolving 136 mg of hypoxanthine and 10 mg of azaserine in 100 ml of sterile distilled water. Sterilize this solution by filtration and store in small aliquots at −20 °C.
- Serum-free DMEM
- DMEM + 20% fetal calf serum (FCS) + HA. Prepare using FCS and HA stocks.
- DMEM containing 2 mM L-glutamine and 100 µg/ml gentamycin
- Dimethylsulfoxide (DMSO)

Method

1. Mix the two cell types in a round-bottom centrifuge tube and spin them down together at 400 g for 5 min at room temperature. (The myeloma cells should have been washed twice with serum-free DMEM to remove all traces of fetal calf serum.)

2. Drain the cell pellet aggressively using an aspirator. Be sure that no medium remains, even to the extent of losing a few cells, then slowly add 0.7 ml polyethylene glycol 1500 solution (pre-warmed and held at 37 °C) during 1 min while resuspending the cells by stirring with the tip of a pipette.

3. Stir for another minute with the end of the pipette.

4. Fill a 10 ml pipette with 10 ml of pre-warmed serum-free DMEM. Slowly drop 1 ml of media on to the cell suspension over the period of 1 min while also mixing with the tip of the pipette. Then add the remaining 9 ml of medium over 3 min again using the pipette tip to mix.

5. Centrifuge the tube at 400 g for about 5 min at room temperature.

6. Gently aspirate off the supernatant, add 5 ml of DMEM containing 20% FCS and HA (pre-warmed and held at 37 °C), and resuspend the pellet very gently using a large-bore pipette.

7. Transfer the cell suspension to 45 ml of DMEM containing 20% FCS and HA, and plate the cells out over five to ten 96-well plates using the multichannel pipette.

8. Put the fusion plates in the CO_2 incubator and try to leave undisturbed

Protocol 9. *Continued*

for ten days. Check once at day three for the absence of contamination and for the presence of massive cell death as the selection starts to take effect. Feed the cells on day three by adding one drop per well of DMEM medium containing 20% FCS and HA.

9. At day 7, replace the medium with fresh DMEM medium containing 20% FCS and HA.

10. At day 10, examine the plates carefully by holding them up to the light and looking at the underside: positive growth of hybridoma cells is indicated by a yellowing of the medium and by the presence of small white spots at the bottom of some wells. Under the microscope large spherical colonies of highly birefringent cells (*Figure 5*) should be visible in 10–30% of the wells.

11. At this stage, take 50 µl samples, either from all the wells or from all the wells containing macroscopic colonies, for testing in the screening assays (see Section 3.2).

12. Repeat the screening test at days 13 and 17.

After screening, all the wells that scored positive may be subcultured by transferring the well contents to a 24-well plate and adding 0.5 ml of fresh medium per well. Cells from the 24-well plate can be used as a source for cloning of the hybridoma line and may also be expanded by subculturing into a 90 mm Petri dish and freezing. If the fusion has been very successful, it is easy to become swamped by trying to subculture and clone too many cells at once. Under these circumstances, it is best to choose just five to ten positives for immediate cloning and freeze the rest for later study. Once the positives have been expanded to the 24-well stage, they should be tested again in the screening assay and also in any alternative, more stringent screen that can be applied. If the fusion resulted in many wells with growth but no ORF-specific antibodies, then the only option is to repeat the fusion under identical conditions. If, however, the fusion failed to produce many colonies at all, then certain alterations to the fusion protocol are worth trying, such as the following:

(a) Feeder layers. Successful fusions do not require feeders but if growth appears poor, then feeder layers may help. We use human diploid MRC5 fibroblasts plated two days earlier at 10^4 cells per well in the 96-well trays.

(b) Media supplements. If the FCS batch is suboptimal then the addition of a cocktail of soluble factors to the medium can be helpful. Make the DMEM medium up as usual with gentamycin and L-glutamine (*Protocol 9, Equipment and reagents*) and then add bovine insulin (0.2 units/ml), pyruvate (0.45 mM), and oxaloacetate (1 mM).

(a)

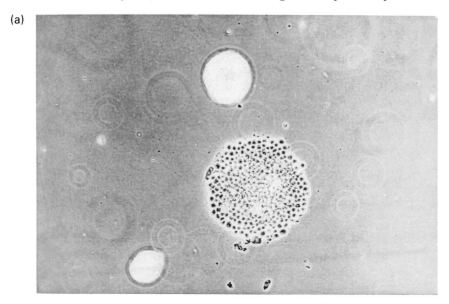

(b)

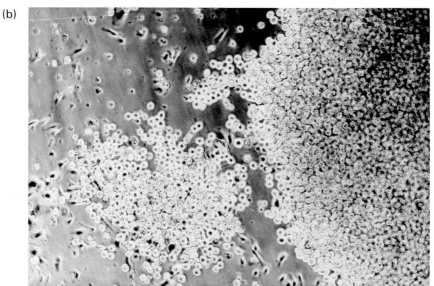

Figure 5. Appearance of growing hybridoma cell cultures. (a) A small colony of cells in a microtitre well. Such colonies are not visible macroscopically and should be present by seven to nine days post-fusion. (b) A large colony, at day 17. The colony is starting to fill the well, the medium is turning yellow, and the cells will need to be subcultured in the next day or two if the well contains antibodies of the desired specificity. (Photographs were kindly provided by Prof. E.B. Lane, Department of Anatomy and Physiology, Dundee University.)

3.8 Freezing hybridoma cell lines

The myeloma parent cells and the hybridoma cells themselves can be safely and efficiently stored in liquid nitrogen freezers for an indefinite period (certainly for at least ten years). The freezing procedure (*Protocol 10*) is straightforward but should be practised and checked using the parent cell before undertaking a fusion. The unusual freezing mix [94% (v/v) FCS, 6% (v/v) DMSO] is essential and mixes that work well for other cell types do not work for hybridomas, so stick to this method.

Protocol 10. Storing hybridoma cell lines [a]

Equipment and reagents

- Liquid nitrogen storage tank
- Freezing mix: 94% (v/v) FCS, 6% (v/v) DMSO
- Sterile freezing vials (Nunc Cryo tubes)

Method

1. Cells for freezing should be growing well so feed them the night before by splitting them 1 in 3 in fresh medium.

2. Spin the cells down at 400 g for 5 min, drain off the supernatant, and gently resuspend in ice-cold freezing mix at the ratio of 0.3 ml of mix to 5×10^6 cells.

3. Aliquot the cells into sterile freezing vials (0.1–0.3 ml per vial), pack the vials in a large polyurethane container filled with cotton wool, and place at −70 °C for 24 h. The vials should then be transferred to the liquid nitrogen container.

4. Ensure that the freezing has a good viability by thawing a test vial a few days later. To avoid accidental losses, always place vials from any given hybridoma in at least two different liquid nitrogen tanks.

[a] Conventional freezing mixes used for freezing other cells do not work well for hybridomas!

3.9 Cloning hybridoma cells

Newly derived hybridomas must be single cell cloned to remove contaminating hybridomas and to select for a cell line which is genetically stable. There are a range of methods available for cloning but the procedure we have found most satisfactory is cloning by limiting dilution. This is detailed in *Protocol 11*.

Protocol 11. Cloning hybridomas by limiting dilution

Equipment and reagents

- Inverted microscope
- 96-Well microtitre plates
- Multichannel pipette and sterile tips

- Tissue culture medium: DMEM containing 20% FCS (and other supplements as required)

Method

1. Set-up a microtitre plate with 50 μl of feeder cells per well (human diploid MRC5 fibroblasts; see Section 3.8) and 50 μl of tissue culture medium per well using a multipipette and sterile tips.

2. Remove 100 μl of hybridoma cell suspension and place in the well at the top left-hand corner (well A1). Mix by pipetting.

3. Serially twofold dilute this cell suspension by transferring 100 μl of cell suspension from the top well to the next well down the column, using a fresh tip for each dilution.

4. Serially twofold dilute the cell suspensions in the left-hand column by transferring 100 μl of cell suspension from the left-most well to the next well on the right, and continue across the plate to the right-hand and side, using the multiwell pipette and a fresh tip for each dilution.

5. Clones should be visible after several days, and should be scored for growth using an inverted microscope. A line running at 45° across the plate should delimit growth from non-growth.

6. Screen wells which are clearly limited by dilution to just one or two clones for antibody at days seven to ten after the initial dilution.

7. Repeat this procedure at least once for clones that produce the antibody of interest.

3.10 Contamination

Scrupulous technique and practice at manipulating the microtitre plates will prevent contamination. It is vital not to overfill the wells, so never have more than 100 μl of medium in a well. If a well is contaminated in an important fusion plate, then *gently* pipette into the well 100 μl of 10 M sodium hydroxide, leave for one hour, then drain the well, wash with 70% ethanol, and leave the well dry. If a well containing antibody secreting cells is contaminated, it may be saved by passage *in vivo*. Inject the entire contents of the well into a BALB/c mouse that has been injected with 200 μl of Freund's complete adjuvant intraperitoneally two to ten days earlier. With luck, an abdominal swelling will appear 10–14 days later as a result of ascites tumour formation. Use a large gauge needle to tap some of the ascites fluid. Place 0.1 ml of fluid in 2 ml of tissue culture medium (DMEM plus supplements and 20% FCS) and examine under the microscope. The hybridoma cells are usually clearly visible and can be cloned immediately (*Protocol 11*). The remainder of the ascites fluid sample can be tested for the presence of specific antibody and should be tested over a wide-range of dilutions as it can be 1000 times more concentrated than tissue culture medium containing secreted antibody.

4. Applications of monoclonal antibodies

Monoclonal antibodies are useful in a wide-range of immunological techniques, some of which are considered in more detail later in this section. They are readily purified using protein A– or protein G–Sepharose chromatography (see *Protocols 12* and *13*) and can be efficiently labelled either directly by iodination (see *Protocol 14*) or indirectly using anti-globulin reagents. Indirect labelling of biotin-coupled antibodies using streptavidin or avidin conjugates is particularly useful if more than one monoclonal antibody is being used in an experiment and we present an easy protocol for their biotinylation (see *Protocol 16*).

Because of their homogeneity and site specificity, monoclonal antibodies are especially useful for immunoaffinity chromatography. We present a powerful method for the preparation of immunoabsorbent columns (see *Protocol 17*) and a set of elution conditions that have allowed us to purify very sensitive, low abundance macromolecules with high efficiency and retention of biochemical activity.

Monoclonal antibodies can be used effectively for immunodetection by immunoprecipitation, immunoblotting, and immunocytochemical procedures. The procedures described in *Protocols 5–7* work well in a range of systems, though the efficacy of an antibody to perform any one of these tests will show individual variation depending upon the antibody–antigen interactions.

The derivation of multiple monoclonal antibodies against the same product allows the establishment of quantitative sandwich immunoassays for the protein using one antibody as a solid phase to 'capture' the antigen from solution, and the other antibody as a labelled 'probe' to detect the immobilized material. Whilst we have used both ^{125}I-labelled and enzyme labelled (usually by biotin labelling the antibody and using an avidin–enzyme conjugate) 'probes' we describe only the former protocol as this has proved more practical in our hands. This assay design can also be used to determine if two antibodies compete for binding to sterically associated sites and also to measure protein–protein interactions.

The full benefits of monoclonal antibodies are realized when the knowledge of their epitope site is combined with the results of their functional effects on the ORF product. This aspect is considered further in Section 4.5.

4.1 Purification of monoclonal antibodies

The purification of mouse IgG antibodies, except those of the IgG1 subclass, is described in *Protocol 12*.

Protocol 12. Purification of antibody subclasses IgG2a, IgG2b, and IgG3

Equipment and reagents

- Spinner vessels
- Protein A–Sepharose column
- 0.1 M citrate buffer pH 3.0: make a 0.1 M solution of citric acid and adjust the pH to 3.0 by adding 0.1 M a solution of trisodium citrate (dihydrated form)
- PBS (*Protocol 4*)
- 1 M Tris–HCl pH 8.5

Method

1. Grow the hybridomas in spinner vessels if possible, using 5% FCS for economy. Do not use adult serum since the immunoglobulins in this will contaminate your preparation.

2. Allow the cells to grow to as high a density as possible by adding extra glucose to the medium (4.5 g/litre) when the cells reach a density of 5×10^5 per/ml. Continue the culture until the medium becomes very acid and cell viability drops below 50%.

3. Spin out the cells and debris (2000 *g* for 15 min at room temperature) and pass the antibody-containing supernatant through a protein A–Sepharose column (5 ml column bed per litre of supernatant) recirculating overnight at a flow rate of 60 ml/h.

4. Wash the column with ten column volumes of PBS and then elute with 0.1 M citrate buffer pH 3.0.

5. Collect the eluted antibody into tubes containing 1 M Tris–HCl pH 8.5 so that it is immediately neutralized.

Protocol 12 works very efficiently for all mouse IgG antibodies except those of the IgG1 subclass. These antibodies have a much lower affinity for protein A and so the method must be modified to obtain a good yield. For this class of antibody, *Protocol 13* should be followed. The purification of other immunoglobulin classes is not considered here since the vast majority of hybridomas isolated using the hyperimmunization and fusion protocols described will be of the IgG class.

Protocol 13. Purification of monoclonal antibody subclass IgG1

Reagents

- 1 M Tris–HCl pH 8.5
- Ammonium sulfate (enzyme grade)
- High salt buffer: 0.1 M Tris–HCl pH 8.5, 3 M NaCl
- Protein A–Sepharose column

Method

1. Adjust the pH of the hybridoma supernatant to pH 8.5 by adding a 0.1 vol. of 1 M Tris–HCl pH 8.5. Then, slowly and with stirring, add solid ammonium sulfate to achieve a 50% saturation at 4 °C and leave stirring overnight.

2. Recover the precipitate by centrifugation at 10 000 *g* for 30 min and resuspend in 1/50 of the starting volume of high salt buffer (0.1 M Tris–HCl pH 8.5, 3 M NaCl). Then dialyse against this buffer overnight.

3. Centrifuge again to clarify the solution and apply it to the protein A–Sepharose column (5 ml column bed per litre of supernatant) pre-equilibrated in the high salt buffer.

4. Wash the column with ten column volumes of high salt buffer and then elute as described for the other IgG subclasses in *Protocol 12*, steps 4 and 5.

4.2 Solid phase radioimmunoassays

If two antibodies are isolated to sterically discrete sites on the ORF, they can readily be used to establish an ORF-specific radioimmunoassay. In these assays, one antibody (the capturing reagent) is immobilized by binding to the plastic wells of a microtitre plate. The other antibody is iodinated to a high specific activity (100 mCi/mg) using the iodogen method described in *Protocol 14*. The solid phase RIA itself is detailed in *Protocol 15*.

These assays do not require pure antigen, have very low backgrounds, and are very reproducible. We have also found them exceptionally valuable for studying protein–protein interactions using pairs of monoclonal antibodies directed to different subunits of multisubunit protein complexes (10).

Protocol 14. Iodination of monoclonal antibodies

Equipment and reagents

- Gamma counter
- Glass tubes (Gallenkamp TES 100 020G 2 × 6 mm soda glass)
- Iodogen (Pierce Chemical Company, 28600)
- Chloroform
- 0.5 M sodium phosphate buffer pH 7.5
- ^{125}I-labelled sodium iodide (Amersham)
- PBS, 1% BSA, 0.01% sodium azide solution
- Sephadex G-50 column

A. *Preparation of iodogen-coated tubes*

1. Dissolve iodogen at 0.5 μg/ml in chloroform.

2. Dispense 100 μl aliquots into glass tubes. If, in retrospect, this gives too fast a rate of iodination then prepare a further set of tubes containing 20 μl per tube.

3. Dry overnight in a fume-hood to evaporate off the chloroform.

4. Store in a desiccator at room temperature; the iodogen-coated tubes last indefinitely.

B. *Iodination reaction*

1. Add 50 μl of antibody (at 1 mg/ml in 0.5 M sodium phosphate buffer, pH 7.5) to an iodogen-coated tube.

2. Add 500 μCi ^{125}I-labelled sodium iodide ($\sim$ 5 μl of stock) and incubate for 2–10 min.

3. Transfer the contents of the tube to 150 μl of PBS containing 1% BSA and 0.01% sodium azide.

4. Fractionate on a 5 ml Sephadex G-50 column using the same buffer. Monitor fractions with a gamma counter and collect the first peak. The second peak contains unincorporated ^{125}I.

5. Store the iodinated antibodies at 4 °C or − 20 °C.

Protocol 15. Solid phase radioimmunoassay

Equipment and reagents

- Gamma counter
- Plastic microtitre plates
- Multichannel pipette
- PBS (*Protocol 4*)
- 'Solid phase' antibody at 20 μg/ml in PBS

- Iodinated antibody (10^6 c.p.m./ml) in PBS containing 3% BSA and 0.1% NP-40
- Blocking solution: 5% BSA in PBS
- Wash solution: PBS, 0.1% NP-40

Method

1. Prepare antibody-coated plates by adding 50 μl of the antibody solution to each well and incubating in a humidified box at room temperature overnight.

2. Shake out the antibody solution and add 50μl of the blocking solution to each well.

3. After 1 h, wash out the wells with two changes of PBS, 0.1% NP-40 and add the antigen-containing solution (the assay has a sensitivity range of 0.1 pM to 10 pM for most protein antigens). Incubate for 2 h at room temperature.

Protocol 15. *Continued*

4. Wash out the antigen with four changes of the wash solution then one wash of PBS alone.

5. Add 50 μl of iodinated antibody (at 10^6 c.p.m./ml in PBS containing 3% BSA and 0.1% NP-40) per well and incubate for another 2 h at room temperature.

6. Finally remove the labelled antibody, wash the plate as before, and dry off in a fume-cupboard or with a hair dryer. Cut out the individual wells with scissors (or a hot wire machine) and count in the gamma counter.

4.3 Biotinylation

Purified monoclonal antibodies are readily modified by the covalent coupling of biotin groups. This simple procedure acts as an alternative to direct labelling with iodine or a fluorochrome and allows detection of the modified antibody with commercially available labelled streptavidin or avidin reagents. The protocol that we follow is detailed in *Protocol 16*. Kits are also available from commercial sources (e.g. Amersham).

Protocol 16. Biotin coupling of monoclonal antibody

Reagents

- 10 mg/ml solution of biotin succinimide ester (Miles Ltd.) in DMSO
- Monoclonal antibody: 1–3 mg/ml in 0.1 M sodium borate buffer (boric acid/sodium tetraborate) or 0.1 M sodium hydrogen carbonate buffer pH 8.8. Dialyse extensively against either buffer to remove any sodium azide previously used to store the antibody.
- 1 M NH$_4$Cl
- Sephadex G-50 column
- PBS containing 1% FCS and 0.01% sodium azide

Method

1. Dialyse a 1–3 mg/ml solution of antibody against an amino-free buffer, such as 0.1 M sodium borate or 0.1 M sodium hydrogen carbonate at pH 8.8.

2. Add the 10 mg/ml biotin succinimide ester solution to a 1–3 mg/ml solution of the antibody at a ratio of 1 mg protein to 25–250 μg ester. Mix well, and incubate at room temperature for 4 h.

3. Add 20 μl of 1 M NH$_4$Cl per μg of protein or per 250 μg of ester.

4. Dialyse the biotinylated antibody or fractionate it on a Sephadex G-50 column to remove the ester.

5. Store at 4 °C in PBS containing 1% FCS and 0.01% sodium azide.

Biotin-coupled monoclonal antibody can also be used for two site immuno-assays using the protocol essentially described in *Protocol 15*. In this situation, the bound antibody is detected with a commercially available streptavidin horse-radish peroxidase conjugate and with a 3,3′,5,5′ tetramethyl benzidine substrate (see *Protocol 4*).

4.4 Antibody affinity columns

Because of their homogeneity and site specificity, monoclonal antibodies are especially useful for immunoaffinity chromatography. We present a powerful method for the preparation of immunoabsorbent columns (*Protocol 17*) that has allowed us to purify very sensitive, low abundance macromolecules with high efficiency and retention of biochemical activity.

Protocol 17. Coupling monoclonal antibody to protein A

Equipment and reagents

- Standard column chromatography equipment
- Rotating wheel, rocker, or shaker
- Protein A–Sepharose beads
- Falcon centrifuge tubes
- 0.1 M borate buffer pH 9.0 (boric acid/sodium tetraborate)
- Dimethyl-pimelimidate dihydrochloride (Sigma). Store in a desiccator at −20°C

- 0.2 M ethanolamine pH 8.0 (pH with HCl)
- Antigen buffer (varies with the antigen): one buffer used is 20 mM Tris–HCl, pH 8.0, 120 mM NaCl, 1 mM EDTA, 0.1% NP-40
- Elution buffer[a]
- Neutralizing buffer: 1 M Tris–HCl pH 8.8
- PBS containing 0.01% (w/v) sodium azide

A. *Preparing the coupled monoclonal antibody–protein A–Sepharose matrix*

1. Pump 100 ml of hybridoma supernatant (the antibody can be IgG2a, IgG2b, or IgG3, but not IgG1) through a 1 ml column of protein A–Sepharose beads overnight at 4°C.

2. Wash the column thoroughly with 10 vol. of 0.1 M borate buffer pH 9.0 at 4°C. Transfer the column contents to a 15 ml Falcon tube. Harvest the beads by centrifugation for 5 min in a centrifuge at 500 *g* and at room temperature.

3. Add 2 ml of 0.1 M borate buffer pH 9.0. Remove a small aliquot (50–100 µl) to check, using SDS–PAGE, that the amount of bound immuno-globulin is 0.5–2.0 mg/ml of beads.

4. Add dimethyl-pimelimidate dihydrochloride to the bead suspension at a final concentration of 20 mM, to cross-link the antibody to the column. (Store the dimethyl-pimelimidate dihydrochloride in a desiccator at −20°C and add it to the bead suspension immediately after weighing out.) Mix thoroughly and leave for 1 h at room temperature, with mixing on a rotating wheel or a rocker/shaker.

Protocol 17. *Continued*

5. Stop the reaction by washing the beads with 0.2 M ethanolamine pH 8.0.

6. Wash thoroughly with PBS containing 0.01% sodium azide and store at 4 °C. They remain stable for at least a year.

B. *Column chromatography*

1. Prepare a 1 ml column of the coupled monoclonal antibody–protein A– Sepharose matrix.

2. Pre-elute the column with the chosen buffer for antigen elution (such as 0.1 M citrate pH 3.0). Check that no protein is released.

3. Equilibrate the column in a suitable buffer for the antigen sample (e.g. 20 mM Tris–HCl pH 8.0, 120 mM NaCl, 1 mM EDTA, 0.1% NP-40).

4. Load the antigen by recirculating the extract through the column and wash extensively with at least 100 ml of buffer.

5. Elute by pumping the elution buffer[a] in a reverse direction of flow to the loading of the antigen. Collect 500 μl of fractions directly into neutralizing buffer (100 μl of 1 M Tris–HCl pH 8.8).

6. After elution, re-equilibrate the column with PBS containing 0.01% sodium azide.

[a] The elution buffer should be determined empirically; try low pH (3.0–1.5), high pH (10.0– 12.5), high salt (3.5 M $MgCl_2$ or 5 M LiCl), 10% dioxane or 25–50% ethanediol, and combinations thereof.

4.5 Epitope mapping

To exploit the full potential of the antibody's specificity in functional assays it is essential to determine the binding site as accurately as possible. One simple and direct approach is to generate unidirectional deletions of the DNA encoding the ORF fusion protein, and then establish by immunoblotting cell extracts or colonies which constructs are recognized by the antibody. Generally, this should narrow down the binding site to around 50 amino acids. Further definition of the antibody binding site can then be determined from an overlapping series of peptides. Alternatively, the antibody can be used to select out its epitope from an epitope library. We have used principally two types of epitope library. One type, the bacteriophage peptide library, has in effect a random peptide inserted into the coat protein of a bacteriophage (11). By successive rounds of biopanning the library with the antibody to amplify the phage that bind strongly, it is possible to obtain clones that contain epitopes related to the original immunogen. The epitope, and conservative variations thereof, can then be determined by sequencing the bacteriophage

DNA into which random oligonucleotide duplexes had been inserted to make the library. With this technique our laboratory has mapped the epitope of several p53 antibodies to five or six amino acids in a short period of time (12). However, the bacteriophage epitope library is not successful with every antibody, even in cases where the antibody is known to recognize a continuous linear epitope. The uses and limitations of such peptide libraries is discussed in ref. 13.

From past experience we would recommend the second type of epitope library which is available in kit form, the Novatope Library Construction System (Novagen or AMS Biotechnology). Initially, the ORF clone DNA of interest is degraded to small fragments by DNase I and then cloned into a plasmid gene encoded T7 gene 10 protein at the C-terminal end. The T7 gene 10 protein into which the small ORF fragments are fused accumulates in the periplasm of the host *E. coli* cell and so is protected from proteolytic degradation. The epitope library is screened by colony immunoblotting and the epitope is again identified by sequencing the inserts of the positive clones. By titrating the amount of DNase I used, and size selection of the DNase I fragments, linear epitopes can be determined down to 10–20 amino acids in a week or so by comparing the DNA sequences of several clones.

Acknowledgements

We thank the Cancer Research Campaign of the United Kingdom for their support. We would also like to thank our colleagues for their kind permission to reproduce their figures. Thanks are also due to Miss Alison Mackay for typing a large part of the manuscript.

References

1. Tabor, S. and Richardson, C. C. (1985). *Proc. Natl. Acad. Sci. USA*, **82**, 1074.
2. Studier, F. W., Rosenberg, A. H., Dunn, J. J., and Dubendorff, J. W. (1990). In *Methods in enzymology* (ed. D. V. Goeddel). Vol. 185, pp. 60–89. Academic Press, San Diego.
3. Sambrook, J., Fritsch, E. E., and Maniatis, T. (ed.). (1989). *Molecular cloning, a laboratory manual* (2nd edn). Cold Spring Harbor Laboratory Press, Cold Spring Harbor, NY.
4. Midgley, C. A., Fisher, C. J., Bartek, J., Vojtesek, B., Lane, D., and Barnes, D. M. (1992). *J. Cell Sci.*, **101**, 183.
5. Vojtesek, B., Bartek, J., Midgley, C. A., and Lane, D. P. (1992). *J. Immunol. Methods*, **151**, 237.
6. Ostrove, S. and Weiss, S. (1990). In *Methods in enzymology* (ed. M. P. Deutscher), Vol. 182, pp. 371–9. Academic Press, San Diego.
7. Smith, D. B. and Johnson, K. S. (1988). *Gene*, **67**, 31.

8. Riggs, P. (1990). *Current protocols in molecular biology.* (ed. F. M. Ausubel, R. Brent, R. E. Kingston, D. D. Moore, J. G. Seidman, J. A. Smith, and K. Struhl). Greene Associates/Wiley Interscience, NY, pp. 16.6.1–16.6.4.
9. Nilsson, B., Abrahmsem, L., and Uhlen, M. (1985). *EMBO J.*, **4**, 1075.
10. Gannon, J. V. and Lane, D. P. (1987). *Nature*, **329**, 456.
11. Scott, J. K. and Smith, G. P. (1990). *Science*, **249**, 386.
12. Stephen, C. W. and Lane, D. P. (1992). *J. Mol. Biol.*, **225**, 577.
13. Lane, D. P. and Stephen, C. W. (1993). *Curr. Opinon Immunol.*, **5**, 268.

5

Expression of cloned genes in yeast

MICHAEL A. ROMANOS, CAROL A. SCORER,
and JEFFREY J. CLARE

1. Introduction

Saccharomyces cerevisiae has long been a model organism for biochemical and genetic studies, and the expression of cloned genes in yeast was examined as soon as transformation procedures became available. The particular advantages it offers over other expression systems are mainly due to the fact that it is both a micro-organism and a eukaryote. Its genetics are more advanced than any other eukaryote, so that it can be genetically manipulated almost as easily and rapidly as *Escherichia coli*. As a eukaryote, it is more likely to provide a suitable environment for the folding of foreign eukaryotic proteins than *E. coli*, and can be more readily used for secretion of proteins. Yeast can be grown rapidly on simple media, and to high cell density. Additionally, as a food organism, yeast is regarded as highly acceptable for the production of pharmaceutical proteins.

Yields of foreign proteins are usually considerably higher in *S. cerevisiae* (typically 1–5% of cell protein) than in mammalian cells, though lower than in *E. coli* or baculovirus. As it is quick to use and simple to scale up, yeast could be the first eukaryotic system to try if *E. coli* is not successful. There is now a wide choice of expression systems based on *S. cerevisiae* or other yeasts; for a detailed coverage of these systems and when they might be used, the reader is referred to a recent comprehensive review (1). In deciding whether to use a yeast expression system, the following points should be considered:

(a) Intracellular proteins which are insoluble in *E. coli* are often soluble when expressed in yeast, or other eukaryotes. Eukaryotic intracellular proteins are almost invariably soluble and active when produced in yeast.

(b) Proteolytic degradation is much less of a problem in yeast than in *E. coli*.

(c) Many post-translational modifications (eg. phosphorylation) that are not carried out in bacteria are carried out in yeast.

(d) Yeast has proved particularly successful in the secretion of small, unglycosylated proteins such as epidermal growth factor (EGF), and may be considered the system of choice for such products. The secretion of large foreign proteins is less predictable, though often successful. However, the production in yeast of mammalian glycoproteins for therapeutic use is not at present feasible due to the differences in glycosylation.

(e) Since it can be rapidly manipulated, yeast is a good eukaryotic system to use for site-directed mutagenesis studies, where rapid expression is required from many variants of a gene.

This chapter will focus on basic strategies and techniques for the high level expression of genes in yeast, although the techniques can be used in related applications, such as screening cDNA expression libraries in yeast, or constructing engineered yeast strains for drug screening. Situations which commonly cause problems with yeast expression systems are highlighted in the last part of the chapter, along with the techniques for diagnosing problems, and for attempting to solve them.

2. S. cerevisiae expression systems

2.1 Introduction

Most yeast expression vectors are based on the 2μ plasmid which is naturally present in most *Saccharomyces* strains. For efficient transcription of a foreign gene, a yeast promoter and terminator are also required, and multiple restriction sites can be placed between these for insertion of foreign genes. Finally, a yeast selectable marker and sequences for selection and propagation in *E. coli* are required. There is now a wide choice of vector systems for propagating foreign DNA in yeast, and of promoter systems, constitutive or tightly regulated. In most cases, multicopy 2μ vectors containing a powerful, regulated promoter are suitable; a few examples of such vectors are shown in *Figure 1*.

Figure 1. *S. cerevisiae* expression vectors. pWYG7L (1), pWYG4 (1), pWYG2L (1), pYcDE2 (2), and pMA91 (3) are multicopy 2μ-based vectors; pBM150 (4) is a low copy number centromere vector. Selection markers are indicated by *shaded* (yeast) or *cross-hatched* (bacterial) *arrows*; promoters by *black arrows*; terminators by *shaded boxes*; 2μ sequences by *open boxes*. Cloning sites for insertion of foreign genes are marked. The promoters used are indicated; for example, pWYG7L uses the *GAL7* promoter (pGAL7). pWYG7L and pWYG2L differ only in the promoter used, derived from the *GAL7* and *ADH2* genes respectively. Initiating ATG codons are present only in cloning site linkers that have *Nco*I sites. Note that pWYG7L, pWYG2L, pWYG4, and pBM150 make use of terminator elements in downstream yeast sequences (2μ plasmid or *URA3*) and these are not shown. The following abbreviations have been used: AMP-R, β-lactamase gene giving rise to ampicillin resistance; *URA3*, *TRP1*, and *LEU2D S. cerevisiae* genes encoding orotidine-5′-phosphate decarboxylase, *N*-(5′-phosphoribosyl) anthranilate isomerase, and β-isopropylmalate dehydrogenase (promoter defective gene), respectively. *ARS1* and *CEN4*, *S. cerevisiae* replication origins and centromere elements, respectively. *KAN-R*, *Tn903* kanamycin-resistance gene.

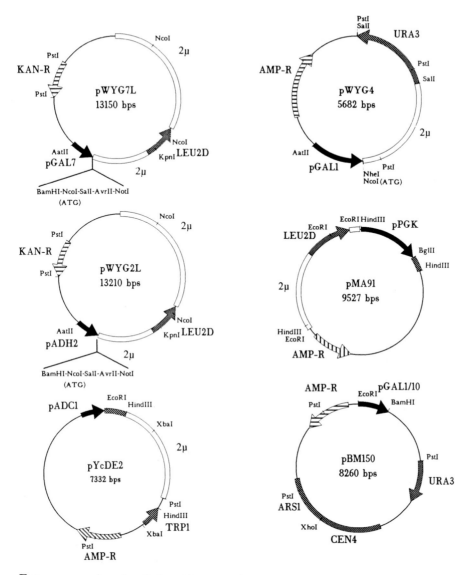

For a comprehensive listing of expression vectors see ref. 5. We will briefly discuss some of the alternative choices available in the components of a yeast expression system, and where these may be useful.

2.2 Vectors for propagation of cloned DNA in yeast

By far the most commonly used vectors for the expression of cloned DNA in yeast are multicopy *E. coli* shuttle vectors based on the yeast plasmid 2μ. However, there are situations when low copy vectors, either episomal or

Table 1. S. cerevisiae vector systems

Vector	Yeast sequences	Transformation frequency [a]	Copy no. per cell	Mitotic instability [b]
Integrating				
YIp	Homologous DNA	10^2	≥ 1	0.1%
Transplacement	Homologous DNA	10	1	Stable
rDNA [c]	rDNA	n/a	100–200	Stable
Episomal				
Replicating (YRp)	ARS	10^4	1–20	20%
Centromere (YCp)	ARS/CEN	10^4	1–2	1%
2μ-based (YEp)	ORI, STB, in 2μ⁺ host (YEp13)	10^4	25	2.8%
	ORI, STB, REP1, REP2, FLP in 2μ⁰ host (pJDB248)	10^4	50–100	0.6–1.8%
	ORI, STB, REP1, REP2, in 2μ⁰ host (pJDB219)	n/a	200	0.26%
	ORI, STB, REP1, REP2, D, FLP in 2μ⁰ host (pJB205) [d]	n/a	50–100	0.20%

[a] Transformants per microgram DNA using the sphaeroplast method.
[b] Plasmid-free cells arising per generation during non-selective growth.
[c] See ref. 6.
[d] See ref. 7.

chromosomally integrated, are required. The different vector systems available are summarized in *Table 1*.

2.2.1 Episomal vectors

i. ARS and ARS/CEN plasmids

DNA can be propagated extrachromosomally either on plasmids containing yeast chromosomal origins of replication (*ARSs*), or on the 2μ plasmid. *ARS* vectors (YRp) are present in 1–20 copies per cell but are mitotically highly unstable. *ARS* vectors can be stabilized by addition of yeast centromeric sequences (*CEN*), but the copy number is then reduced to one or two per cell. In practice, *ARS* vectors are hardly ever used for foreign gene expression. However, *ARS/CEN* vectors (YCp; e.g. pBM150, *Figure 1*) are convenient to use when low level expression is desired (see ref. 5 for examples).

ii. 2μ-based vectors

2μ is a 6.3 kb plasmid present in most *Saccharomyces* strains at about 100 copies per cell (*Figure 2*.) The 2μ plasmid utilizes two mechanisms to achieve stable maintenance:

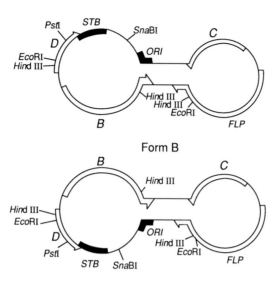

Figure 2. The two forms of the 2μ plasmid. *Cis* elements are shown as *filled boxes (ORI,* replication origin; *STB,* stability locus) and genes by *open boxes (FLP, B, C,* and *D* genes), inverted repeat regions are aligned; see text for a detailed description. Reprinted by permission of John Wiley and Sons Ltd., from ref. 1.

(a) Amplification to correct fluctuation in copy number caused by inefficient segregation. This depends on the *FLP* gene product, which is a site-specific recombinase catalysing inversion about two inverted repeats, resulting in two forms of the plasmid, A and B (*Figure 2*).

(b) A system for reducing the maternal cell bias in plasmid transmission during budding, increasing the efficiency of segregation. This depends on having the *STB* locus in *cis* and the *REP1* and *REP2* gene products in *trans*.

The simplest 2μ vectors contain the 2μ *ORI-STB* (e.g. YEp13, *Table 1*; pWYG4 is also of this type, *Figure 1*), and are used in a 2μ⁺ host strain which supplies REP1 and REP2 proteins. *ORI-STB* vectors are convenient to use due to their small size and ease of manipulation. They are tenfold more stable than *ARS* vectors, being lost in 1–3% of cells per generation in non-selective conditions, and are present in 10–40 copies per cell. In 2μ-free (2μ⁰) cells such plasmids behave as *ARS* vectors. Commonly used *ORI-STB* vectors contain a 2.2 kb *Eco*RI fragment or a 2.1 kb *Hind*III fragment from the B form of 2μ, each having one inverted repeat. It should be noted that shuttle vectors containing inverted repeats will exist as a variety of recombinants with native 2μ.

More complex 2μ-based vectors contain the *REP1* and *REP2* genes in addition to *ORI-STB* and can therefore be used in 2μ⁰ strains (see *Table 1*). Being larger, they are more difficult to manipulate than *ORI-STB* vectors but are stabler and more suitable for scale up. Many examples of this type of vector are disrupted in *D* (e.g. pJDB248), and some in *FLP* (e.g. pJDB219). Recently, vectors have been developed which have all the functional regions of 2μ intact, by insertion of the foreign DNA at the unique *Sna*BI site. These are even more stable, though still 20- to 80-fold less stable than 2μ.

A number of vectors (e.g. pWYG7L or pWYG2L, *Figure 1*) are based on pJDB219, and use the *LEU2*-d marker which results in very high copy numbers (200–400 per cell). pJDB219 is best used in 2μ⁰ cells since it can undergo FLP-mediated recombination with resident 2μ, leading to loss of the foreign DNA but retention of the *LEU2*-d marker. In 2μ⁰ cells it is very stable due to its high copy number, making it very suitable for large scale culture.

2.2.2 Integrating vectors

i. *Single cross-over integration*

Chromosomal integration, which in *Saccharomyces* occurs by homologous recombination, offers a more stable alternative to episomal maintenance of foreign DNA (*Figure 3*). Integrating vectors (YIp) contain yeast chromosomal DNA to target integration, as well as a selectable marker, and bacterial replicon. Vectors are digested at a unique restriction site in the homologous DNA to promote high efficiency integration; for convenience the selection marker is usually used to target integration. Though the resulting integrants have a duplication of the target sequence, so that the vector can potentially 'pop-out' by recombination, they are nevertheless relatively stable. If high concentrations of integrating vectors are used in transformations, tandem multicopy inserts can result, due to repeated recombination events. Recently, multicopy expression vectors have been developed based on integration into reiterated chromosomal DNA, e.g. ribosomal DNA (see *Table 1*).

ii. *Transplacement*

An alternative type of integration, transplacement, makes use of double homologous recombination to replace chromosomal DNA, resulting in a

---

Figure 3. Chromosomal integration of DNA by homologous recombination. (a) Targeted single cross-over integration. The vector is linearized by digestion at a unique restriction site (A) within the *URA3* gene, targeting integration into the chromosomal *ura3* locus by single cross-over recombination. (b) Transplacement. A transplacing fragment, consisting of the expression cassette and selectable marker (*URA3*) flanked by DNA homologous to the 5' and 3' ends of the target chromosomal DNA, is released from the vector by digesting with restriction enzyme B. After transformation, this fragment integrates into the chromosome by homologous double cross-over recombination resulting in replacement of the chromosomal DNA.

a)

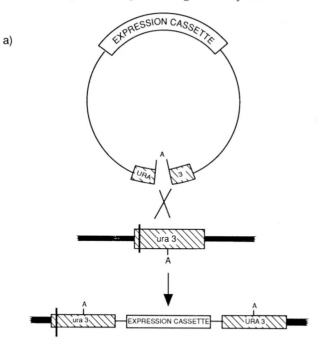

b)

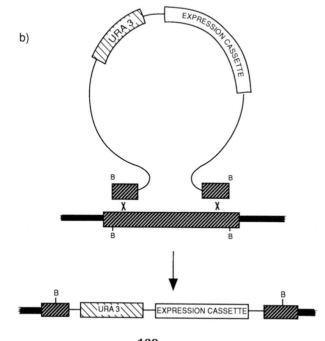

stable structure without duplications (*Figure 3*). When a stable single copy transformant is required, this is the method of choice. However, transformants should be checked by Southern blot analysis to confirm the predicted transplacement event. (In contrast, episomal YCp vectors can be used without analysis.) Transplacement vectors contain the exogenous DNA and selection marker flanked by yeast DNA homologous to 5' and 3' regions of the chromosomal DNA to be replaced. Prior to transformation the vector is digested with restriction enzymes which liberate the transplacing fragment with 5' and 3' homologous ends.

2.3 Selection markers

Selectable markers are required both for the isolation and continued selection of transformants; a wide variety is now available, some are listed in *Table 2*. The first and most commonly used were *LEU2*, *TRP1*, *URA3*, and *HIS3*, used in strains auxotrophic for leucine, tryptophan, uracil, and histidine, respectively. Such strains are widely available, and many contain non-reverting mutant alleles or deletions constructed to give low background rates in transformations (*Table 4*). A variant of *LEU2*, *LEU2*-d, has a truncated promoter and is poorly expressed, so that its selection gives rise to very high plasmid copy numbers.

Dominant selectable markers increase the range of host strains that can be used, to include prototrophic and industrial strains of *S. cerevisiae*, and can be used for selection in rich medium. These properties are valuable when analysing

---

Table 2. Selectable markers for *S. cerevisiae* transformation

Marker	Auxotrophic/ dominant	Comments
HIS3	A	—
TRP1	A	Selection possible in casamino acids
LEU2	A	*LEU2*-d for high copy number selection
URA3	A	• Selection possible in casamino acids
		• *URA3*-d for high copy number selection
		• Counter-selection using 5-fluoro-orotic acid
		• Autoselection in *fur1* (5-fluorouracil-resistant) strain
LYS2	A	Counter-selection using α-amino adipate
Tn903 kan[r]	D	Active only in multiple copies unless yeast promoter used; selection using G418
Cm[r]	D	Only effective using yeast promoter; selection using chloramphenicol in glycerol medium only
Hyg[r]	D	—
CUP1	D	Level of Cu^{2+}-resistance dependent on gene dosage
DHFR	D	Methotrexate/sulfanilamide selection; level of resistance dependent on gene dosage

expression in a range of strains in order to optimize yield, or in mutant strains (e.g. glycosylation mutants) which are not multiply marked. The vectors pWYG7L and pWYG2L (*Figure 1*) contain the *Tn903 kan'* gene, which confers resistance to the aminoglycoside antibiotic kanamycin in *E. coli* or to G418 in yeast.

2.4 Transcriptional promoters and terminators

In general, foreign promoters are not functional in yeast and it is therefore necessary to use a yeast promoter for efficient expression of heterologous genes. Most yeast promoters are regulated to some extent, but the most powerful (those from genes encoding glycolytic enzymes) are poorly regulated. As with any other expression system, it is desirable to use a tightly regulated promoter in order to separate the growth phase of the host cells from the induction. This minimizes the metabolic burden on the host cell and avoids the selection of low expressing variants. The use of regulated systems is particularly desirable in large scale culture and is essential when expressing toxic proteins. A variety of promoter systems have been used for expression of cloned genes, some are listed in *Table 3*.

Table 3. *S. cerevisiae* promoter systems

Promoter	Strength	Regulation[a]	Reference
Native			
PGK, GAP, TPI	++++	≥ 20-fold induction by glucose (PGK)	8, 9
GAL1	+++	1000-fold induction by galactose	10
ADH2	++	100-fold repression by glucose	11
PHO5	+/++	200-fold repression by P_i	12
CUP1	+	20-fold induction by Cu^{2+}	13
MFα1	+	Constitutive in α cells	14
MFα1	+	10^5-fold induction by shift 37 °C to 24 °C in strain JRY 188[b]	14
Engineered			
PGK/α2 operator	++++	100-fold induction by shift 37 °C to 24 °C in sir3[ts] host	15
GAP/GAL	+++?	150- to 200-fold induction by galactose	16
PGK/GAL	+++?	Induction by galactose	17
GAP/ADH2	+++?	Repressed by glucose	18
CYC1/GRE[c]	+++?	50- to 100-fold by deoxycorticosterone in host strain expressing glucocorticoid receptor	19
PGK/ARE[d]	++++	Several 100-fold by dihydrotestosterone in host strain expressing androgen receptor	20

[a] Induction ratios are reporter-dependent.
[b] See *Table 4*.
[c] Glucocorticoid responsive element.
[d] Androgen responsive element.

The first promoters used for foreign gene expression in yeast were from genes encoding abundant glycolytic enzymes, e.g. alcohol dehydrogenase I (*ADH1*), phosphoglycerate kinase (*PGK*), or glyceraldehyde-3-phosphate dehydrogenase (*GAP*). These were initially thought to be constitutive but were later found to be induced several-fold by glucose. Glycolytic promoters are the most powerful of *S. cerevisiae*, and have been used extensively with many successes despite their poor induction ratio. The most commonly used regulated promoters are those of the *GAL1*, *GAL7*, and *GAL10* genes involved in galactose-metabolism; they are induced over 1000-fold by the addition of galactose, and are repressed by glucose. Another powerful tightly regulated promoter that has been used to express many foreign genes is that of the glucose-repressed *ADH2* gene. There is now also a variety of engineered hybrid promoters, most of which have been constructed in an attempt to combine tight regulation with the strength of glycolytic promoters (*Table 3*).

The process of mRNA 3' end formation ('termination') in yeast is similar to that of higher eukaryotes in involving three steps: termination of transcription, endonucleolytic processing, and polyadenylation. However, neither polyadenylation/termination signals from higher eukaryotes, nor prokaryotic terminators, are generally functional in yeast. Since efficient termination is probably required for maximal gene expression, yeast 'terminators' are usually included in expression vectors. Terminators from a number of genes have been used, but in order to simplify vector construction, a terminator from 2μ is often used, e.g. in pWYG7L and pWYG4 (*Figure 1*).

2.5 Engineering a foreign gene for intracellular expression

Having decided to express a gene in yeast, the first step is to tailor it so as to maximize the possibility of efficient transcription and translation. Since introns from higher eukaryotes are not accurately spliced in yeast, only intronless genes or cDNAs can be expressed. In adapting the foreign gene for expression in yeast it is advisable to carry out the following:

(a) Eliminate all non-coding sequence especially in the 5' untranslated region (UTR), since upstream ATGs, high G content, and secondary structure in the 5' UTR are all highly inhibitory to translational initiation in yeast.

(b) Precede the initiation codon by AT-rich DNA, preferably AAAAAATG, since this favours efficient translational initiation. (The A at position −3 is the most important, though the ATG context is much less important than it is in higher eukaryotes, where the preferred context is CACC*ATG*G.)

(c) Include suitable restriction sites for direct cloning into the expression vector. Here it is worth selecting sites that will allow:

- transfer to different systems, e.g. baculovirus (which has similar upstream sequence requirements; vectors often have a *Bam*HI cloning site) or *E. coli* (this may be difficult because of the need for a Shine–Dalgarno sequence, but vectors with a *Nco*I site at the initiating ATG can be made compatible)

- transfer to secretion vectors (e.g. using a *Nco*I site for transfer to pWYG82, see Section 3.3.1)

- expression of fusion proteins.

The simplest method for cloning the foreign gene is to use synthetic oligo-nucleotides to reconstruct the 5′ and 3′ ends with the adapted sequence. Alternatively, PCR may be used for direct cloning of the adapted gene, although in this case the sequence of the amplified DNA must be verified. In the ideal situation, all untranslated sequence is removed from the gene apart from the 5′ and 3′ cloning sites and a run of As preceding the initiation codon. Most of the untranslated sequence in the mRNA is then derived from the vector. However, if this is not convenient, it is worth testing an unoptimized construct provided the 5′ UTR lacks ATGs, runs of G, or a predicted RNA secondary structure of $\Delta G > -20$ kcal/mol.

3. *S. cerevisiae* secretion systems

3.1 Introduction

Yeast cells naturally secrete few proteins. However a wide variety of foreign proteins have been secreted efficiently from yeast. Secretion offers the following advantages over intracellular expression:

- Many naturally secreted proteins (e.g. human serum albumin, HSA) can adopt their correct conformation and disulfide bonding only by folding in the secretory pathway. For such proteins, secretion is the only option.

- Foreign proteins secreted to a high level may need only minimal purification from the culture medium.

- Secretion of intracellular proteins is sometimes possible and can be used to express proteins which are toxic or unstable in the cytoplasm (e.g. HIV protease; ref. 21).

The secretion of foreign proteins has been most successful with small polypeptides (e.g. EGF or insulin) or unglycosylated proteins such as HSA. In high density fermentations, these products can be secreted in some cases to levels in excess of one gram per litre. The secretion of glycoproteins is problematic, however, since yeast *N*- and *O*-linked carbohydrate differs from that in mammalian cells. Since yeast-derived glycoproteins are antigenic (22), the production of therapeutic glycoproteins in yeast is not feasible at present.

3.2 Yeast signal sequences

A prerequisite for the secretion of most proteins is the presence of an N-terminal hydrophobic signal peptide (approximately 20 residues) which is cleaved in the secretory pathway. Most heterologous signal peptides will not direct secretion efficiently in yeast and therefore secretion vectors usually contain a yeast signal sequence, usually from the yeast mating pheromone α-factor, invertase, or acid phosphatase (encoded by the *MFα1*, *SUC2*, and *PHO5* genes, respectively). Of these the α-factor leader (prepro region) has been used most. *MFα1* encodes a 165 amino acid protein, prepro-α-factor, which comprises a signal peptide and a pro region, followed by four tandem repeats of the mature α-factor. Each repeat is preceded by a short spacer peptide (Lys–Arg–(Glu/Asp–Ala)$_{2-3}$). Processing involves four proteolytic steps (1):

- cleavage of the signal peptide
- cleavage after Lys–Arg by KEX2 protease
- removal of Glu–Ala pairs by STE13 diaminopeptidase
- removal of C-terminal Lys and Arg residues by KEX1 protease.

These steps, shown diagrammatically in *Figure 4*, lead to the secretion of four molecules of α-factor.

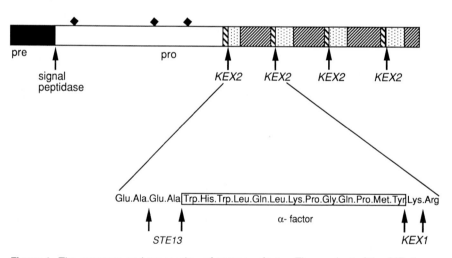

Figure 4. The structure and processing of prepro-α-factor. The product of the *MFα1* gene is shown schematically. The three *N*-linked glycosylation sites in the pro region are marked ◆. The peptide product from the cleavage of prepro-α-factor by KEX2 is also indicated and sites for further processing by STE13 and KEX1 are shown. Reprinted by permission of John Wiley and Sons Ltd., from ref. 1.

3.3 Engineering foreign genes for secretion vectors

Secretion vectors express fusion proteins containing a yeast signal peptide joined to the foreign protein (without its signal peptide). These fusions must be designed to allow efficient and accurate processing of the chimaeric pre-protein, resulting in a secreted foreign protein with the correct mature N-terminus.

3.3.1 α-Factor secretion vectors

DNA encoding the mature foreign protein (i.e. without its own signal) is fused in-frame after the α-factor prepro DNA. This is generally achieved by using an engineered restriction site in *MFα1*, 5' to the DNA encoding the cleavage site; the gene can then be ligated to this site using a synthetic linker that spans the junction, or by PCR. This strategy is illustrated using the α-factor secretion vector pWYG82 in *Figure 5*.

With multicopy vectors, removal of the Glu–Ala spacers is rate limiting, so that the secreted protein is a mixture of mature protein and protein containing N-terminal Glu–Ala spacers. Therefore the Glu–Ala spacers are often omitted from the α-factor fusion, so that cleavage after Lys–Arg generates the mature protein (*Figure 5*). This strategy was first used successfully with EGF (14), but in many other cases removal of the Glu–Ala spacers reduces the efficiency of cleavage after Lys–Arg, so that proprotein accumulates. A reasonable strategy is to make secretion vectors with the spacer peptide in the first instance. If this gives efficient secretion, a vector without the spacer can then be tested. The secretion vector pWYG82 can be conveniently used for the production of proteins with or without the Glu–Ala spacers.

It should be noted that some α-factor secretion vectors use the *MFα1* promoter, which is constitutive but only active in cells of α mating type. These vectors can also be used for temperature-regulated secretion of proteins (14), in α cells having the *sir3*[ts] mutation (e.g. strain JRY188, see later, *Table 4*).

3.3.2 *SUC2* secretion vectors

Unlike α-factor, invertase has a classical signal peptide that is processed only by signal peptidase (23). There are several examples of invertase/foreign protein fusions where the signal peptide is accurately and efficiently processed, suggesting that processing is not greatly affected by the amino acid after the cleavage site (24). Therefore precise invertase fusions should be processed efficiently to the correct mature product. In comparison to the α-factor system, this has the advantage that there should not be alternative processed forms of the product in the medium (however, the number of published examples using *SUC2* fusions is far smaller). Vectors with convenient restriction sites in the *SUC2* signal peptide DNA are not readily available. Therefore, since the *SUC2* sequence is relatively short, it can be synthesized and ligated to each foreign gene before cloning into an expression vector.

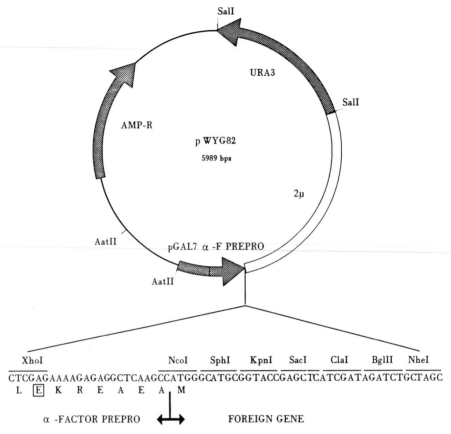

Figure 5. The α-factor secretion vector pWYG82. Foreign genes may be inserted between the *Xho*I site and the other unique sites in the polylinker, using a synthetic oligonucleotide to reconstruct the junction (i.e. the Lys–Arg cleavage site with or without the Glu–Ala spacer). Engineering of the *Xho*I site results in an amino acid substitution adjacent to the KEX2 cleavage site (Asp to Glu, indicated by the box), but this has no effect on secretion efficiency. *URA3* and AMP-R are as defined in the legend to *Figure 1*.

4. Techniques for culturing and transforming yeast

4.1 Culture and storage of yeast strains

Although hundreds of strains of *S. cerevisiae* are available, a small number is sufficient for most procedures described here. Readily available strains are listed in *Table 4*. We have found the strain S150-2B to be the most efficient for galactose inductions; the $2\mu^0$ version should ideally be used with complex 2μ vectors such as pWYG7L, but only $2\mu^+$ strains should be used for *ORI-STB* vectors such as pWYG4 (see Section 2.2.1). It should be noted that many

Table 4. *Saccharomyces cerevisiae* strains commonly used for gene expression

Strain	Genotype	Source/comments
S150–2B	a leu2–113,112 ura3–52 trp1–289 his3Δ1	2μ[+] and 2μ[0] available from A. Cashmore[a]
MC16	a leu2–3 his4–712 SUP2 ade2–1 lys2–1	NCYC[b]
MD40–4C	a leu2–3, 112 his3–11,15 ura2 trp1	NCYC[b]
YNN281	a trp1 his3Δ200 ura3–52 lys2–801 ade2–1	YGSC[c]
KY117	a ura3–52 trp1Δ1 lys2–801 ade2–101 his3Δ200	Ref. 25
AH22	a leu2–3, 112 his4–519, can1	ATCC[d], NCYC[b]
DC04a	a ade1 leu2 [2μ[0]]	YGSC[c]
LL20	a leu2–3,112 his3–11,15	ATCC[d], NCYC[b]; 2μ[0] derivative also available
GM3C-2	a leu2–3,112 trp1-1 his4–519 cyc1–1 cyp3–1	ATCC[d]
JRY188	α leu2–3,112 ura3–52 his3 trp1 sir3[ts]	J. Rine[e]; temperature regulation of MFα1 promoter
BJ5457	a leu2Δ1 ura3–52 trp1 lys2–801 his3Δ200 pep4::HIS3 prb1Δ1 can1 GAL	YGSC[c]; non-revertible proteinase mutant
AA274	a leu2–3,112 ade2 his3Δ200 lys2Δ201 ura3–52 pmr1–1::LEU2	Ref. 26; supersecretor
LB347–1C	α mnn9 mal mel gal2 CUP1	Ref. 27; mutant defective in outer chain glycosylation
XW451–3B	α mnn1 mal mel gal2 CUP1	Ref. 28; mutant lacking antigenic α1, 3-linked mannose
LB 361–1C	a mnn1 mnn9 mal mel gal2 CUP1	Ref. 27

[a] A. Cashmore, Department of Genetics, University of Leicester, Leicester LE1 7RH, UK.
[b] National Collection of Yeast Cultures, AFRC Institute of Food Research, Colney Lane, Norwich NR4 7UA, UK.
[c] Yeast Genetic Stock Center, Department of Molecular and Cellular Biology, University of California, Berkeley, CA 94720, USA.
[d] American Type Culture Collection, 12301 Parklawn Drive, Rockville, Maryland 20852–1776, USA.
[e] J. Rine, Department of Molecular and Cellular Biology, Division of Genetics, 401 Barker Hall, University of California, Berkeley, CA 94705, USA.

commonly used strains have mutations in the galactose permease gene (*GAL2*) and may not be fully galactose inducible. The use of protease-deficient strains (e.g. *pep4* mutants such as BJ5457, *Table 4*) may increase yields of some proteins.

Strains should be stored as frozen glycerol stocks in the long-term, and grown on agar plates as working stocks which can be kept for several months. Untransformed strains should be grown in rich medium (YPD broth or YPD agar; see *Table 5*); transformants should generally be grown in selective medium. Transformants containing the vector pWYG7L (*Figure 1*) can be selected by growth in rich media containing G418 (e.g. YPD + G418). With vectors containing auxotrophic selection markers, minimal medium containing all the necessary supplements should be used (e.g. YNBD + histidine +

Table 5. Commonly used yeast growth media[a]

Rich media

YP	Dissolve 10 g yeast extract (Difco 0127–01–7) and 20 g Bacto peptone (Difco 0118–01–8) in 1 litre distilled water.[b] Autoclave.
YPD	Add 0.1 vol. 20% glucose (autoclaved or filter sterilized) to autoclaved YP.
G418 media	Make a 100 × stock solution (50 mg/ml) of G418 (Geneticin, Gibco 066–01811) in water. Filter sterilize, store at −20 °C. Add to autoclaved media.

Minimal media

YNB	Make up 10 × YNB by dissolving 6.7 g yeast nitrogen base without amino acids (Difco 0919–15–3) in 100 ml of distilled water. Filter sterilize or autoclave. Dilute to 1 × YNB with sterile water.
YNBCas	Prepare a 10 × casamino acids stock (low salt, Difco 0230–01–1) by dissolving 20 g in 100 ml of water and autoclaving. Add 0.1 vol. to YNB.
YNBD	YNB containing 2% glucose.
YNBDCas	YNBCas containing 2% glucose.
Supplements for YNB	Adenine sulfate, uracil, L-tryptophan, L-histidine–HCl, L-arginine–HCl, L-methionine (each at 20 mg/litre). L-Tyrosine, L-leucine, L-isoleucine, L-lysine–HCl (each at 30 mg/litre). L-Phenylalanine (50 mg/litre), L-glutamic acid (100 mg/litre), L-aspartic acid (100 mg/litre), L-valine (150 mg/litre), L-threonine (200 mg/litre), L-serine (375 mg/litre). Make up 100 × stocks and filter sterilize, add to autoclaved media.

[a] To make agar media, add 2% agar to all the medium components except the sugar and autoclave. Before pouring plates, melt the agar medium, add the sugar, and mix.
[b] For most purposes double distilled water is adequate. Occasionally, erratic growth rates are seen, in which case it is advisable to use tissue culture grade water, and to grow cultures only in disposable plastic containers (e.g. Universals or Erlenmeyer flasks).

leucine + tryptophan for the vector pWYG4 (*Figure 1*) in strain S150-2B). The compositions of commonly used growth media are given in *Table 5*. Procedures for culturing and storing yeast are given in *Protocol 1*. Since contaminating moulds grow readily on yeast media, it may be necessary to carry out manipulations in a sterile flow hood.

Cell density in liquid culture may be monitored using a haemocytometer, or more conveniently by measuring A_{600}. An A_{600} of 1.0 is roughly equivalent to 2×10^7 cells/ml; this will vary somewhat depending on the spectrophotometer and should ideally be calibrated. In YPD broth, most strains grow with doubling times of about two hours, to a density of $1–2 \times 10^8$ cells/ml ($A_{600} = 5–10$). The growth rate and final culture density are significantly lower in YNBD, but can be increased, for *URA3* or *TRP1* selection, by addition of casamino acids (since this preparation contains all the amino acids except for tryptophan and lacks uracil).

Protocol 1. Culturing and storing yeast

Equipment and reagents

- Sterile plastic 25 ml Universal container (e.g. Nunc 3-64211)
- Sterile screw-cap vials (1–2 ml)
- YPD, YPD agar, YNBD, and YNBD agar (see *Table 5*)

- 75% glycerol in water, autoclaved or filter sterilized
- 100 × G418 (50 mg/ml, Gibco 066-01811), filter sterilized

Method

1. Streak out the culture on a plate containing YPD agar.[a] Incubate at 30 °C for three days until small colonies appear. Store at 4 °C. Use this plate as a working stock.[b]

2. Inoculate 10 ml of YPD in a 25 ml Universal container using a single isolated colony. Grow for one to two days at 30 °C, shaking at 300 r.p.m. This culture can be used as a starter for preparing exponentially growing cultures for transformation, or for frozen stocks.

3. Make a frozen stock for long-term storage of the culture: place 0.8 ml of culture in a sterile screw-cap vial and add 0.2 ml of 75% glycerol (final glycerol concentration of 15%). Cool as slowly as possible to −70 °C; this is readily achieved by placing in an expanded polystyrene box in a − 70 °C freezer. Store at −70 °C. Cultures remain viable for several years.

4. To use frozen stocks, scrape off a small amount of the frozen culture (do not thaw) using a sterile spatula and streak on to YPD agar.

[a] For transformants containing G418 vectors, use YPD + G418 liquidor solid media throughout. For transformants requiring auxotrophic selection, use YNB + supplements (see *Table 5*).
[b] YPD agar allows longest storage (several months at 4 °C if sealed). Use selective medium for transformants.

It should be noted that constitutive expression vectors may result in reduced growth rate and reduced viability on storage, causing selection for low-expressing variants. Therefore it is particularly important with such systems, and generally good practice, to establish frozen stocks of transformants at once and to minimize subculturing.

4.2 Transformation of yeast

The first method for transforming yeast involved enzymatic removal of the cell wall to generate sphaeroplasts which can take up DNA in the presence of

Ca^{2+} and polyethylene glycol. This method is highly efficient, but is tedious and suffers from the disadvantage that the transformed sphaeroplasts must be regenerated by embedding them in isotonic agar and cannot therefore be replica plated directly. Much simpler techniques involving intact cells are now available which can give almost as high transformation frequencies. Episomal vectors are readily introduced into intact cells by treatment with lithium ions (*Protocol 2*) or by electroporation (*Protocol 3*). These methods can also be used for integrating and transplacement vectors, though transformation frequencies will be reduced.

Protocol 2. Transformation of *S. cerevisiae* using the lithium method

Equipment and reagents

- Sterile 250 ml Erlenmeyer flask (Corning 25600–250)
- Sterile 50 ml centrifuge tubes (e.g. Falcon 2098)
- Sterile 15 ml polypropylene snap-top tubes (e.g. Falcon 2059)
- YPD (see *Table 5*)
- TE buffer: 10 mM Tris–HCl pH 7.4, 1 mM EDTA; filter sterilize or autoclave
- LA buffer: 0.1 M lithium acetate in TE buffer; filter sterilize or autoclave
- 50% polyethylene glycol (PEG) (polyethylene glycol 4000, BDH 44273) 4x: dissolve in TE, then filter sterilize. (Store for one week only)
- Salmon sperm DNA (dissolve in water to 10 mg/ml, shear by syringing, and boil for 10 min; store at $-20\,°C$)
- Plates containing selective medium (YPD agar + G418 or YNBD agar + supplements; see *Table 5*)

Method

1. Prepare an overnight starter culture in YPD of the strain to be transformed. Measure the A_{600}.

2. Dilute the starter culture into 50 ml of YPD in a 250 ml Erlenmeyer flask to an $A_{600} = 0.25$ (typically a 20-fold dilution). Incubate at 30 °C with shaking until the culture density reaches $A_{600} = 1.0$ (4–5 h).

3. Harvest the cells by centrifugation (20 °C, 2000 g, 5 min). Resuspend the pellet in 50 ml TE buffer and centrifuge again.

4. Resuspend the cells in 2.5 ml LA buffer and incubate at 30 °C with shaking for 1 h. This volume of competent cells is sufficient for eight transformations.

5. For each transformation, dispense 0.3 ml of competent cells into a 15 ml snap-top tube. Add up to 10 µg of vector DNA[a] (maximum 20 µl) and 0.7 ml of 50% PEG, mix gently.[b] Incubate at 30 °C for 1 h, without shaking.

6. Heat shock the cells for 5 min at 42 °C.

7. Pellet the cells by centrifugation (20 °C, 2000 g, 5 min) and remove excess PEG.

---

8. Plate out the cells:

(a) If using G418 selection, resuspend the cells in 0.5 ml YPD and incubate overnight at room temperature to allow expression of antibiotic resistance. Pellet the cells by centrifugation, resuspend in 200 μl of TE buffer, and spread on to a single YPD + G418 agar plate.[c] Incubate at 30 °C. Transformants should appear after three days.

(b) If using auxotrophic selection,[d] resuspend the cells in 200 μl of TE buffer and plate out on minimal selective agar (YNBD agar + supplements). Incubate at 30 °C. Transformants should appear after three to seven days

---

[a] Mini-prep DNA may be used.

[b] Transformation frequencies can be increased by adding 200 μg of single-stranded carrier DNA, but this is not usually necessary.

[c] Some strains may give a high background of spontaneous G418-resistant mutants. This problem can be eliminated by selecting on plates containing 2% glycerol instead of glucose. However, colonies will take at least five days to grow.

[d] If selecting for the *LEU2*-d marker, it is necessary to incubate transformants overnight in YPD before selection on minimal medium.

Protocol 3. Transformation of *S. cerevisiae* by electroporation

Equipment and reagents

- Sterile 500 ml Erlenmeyer flask (Corning 25600–500)
- Sterile 50 ml centrifuge tubes (e.g. Falcon 2098)
- Sterile polypropylene snap-top tubes (e.g. Falcon, 2059)
- Sterile 1.5 ml microcentrifuge tubes (autoclaved)
- Electroporator (e.g. Gene Pulser with Pulse Controller, Bio-Rad)
- Electroporation cuvettes, 0.2 cm (cool on ice)

- YPD (see *Table 5*)
- Ice-cold sterile distilled water
- Ice-cold sterile 1 M sorbitol (autoclave or filter sterilize)
- YPD containing 1 M sorbitol (dissolve sorbitol in YPD before autoclaving)
- Plates containing selective medium (YPD agar + G418 or YNBD agar + supplements); ideally use YNBD plates containing 1 M sorbitol (add solid sorbitol to YNBD agar before autoclaving)

Method

1. Prepare an overnight starter culture in YPD of the strain to be transformed. Measure the A_{600}.

2. Dilute the starter culture into 100 ml of YPD in a 500 ml Erlenmeyer flask to an $A_{600} = 0.25$ (typically a 20-fold dilution). Incubate at 30 °C with shaking until the culture density reaches $A_{600} = 1.3$–1.5 (5–6 h).

3. Harvest the cells by centrifugation (4 °C, 2000 g, 5 min). Resuspend the pellet in 50 ml of ice-cold water and centrifuge again.

Protocol 3. *Continued*

4. Resuspend the pellet in 4 ml of ice-cold 1 M sorbitol and centrifuge (4 °C, 2000 *g*, 5 min).

5. Resuspend the pellet in 0.2 ml of ice-cold 1 M sorbitol. This volume of electrocompetent cells (approx. 0.3 ml) is sufficient for seven transformations.

6. For each electroporation, dispense 40 µl of competent cells into a sterile 1.5 ml microcentrifuge tube. Add up to 100 ng of plasmid DNA (maximum 5 µl, in buffer of low ionic strength) and mix gently. Leave on ice for 5 min.

7. Transfer the suspension to a pre-chilled electroporation cuvette taking care not to introduce air bubbles. Tap the cuvette to ensure that the cells cover the base of the cuvette. Pulse at 25 µF, 1.5 kV, 200 Ohms using a Bio-Rad Gene Pulser. The time constant should be approximately 5 msec.

8. Immediately after the pulse, add 1 ml of ice-cold 1 M sorbitol and mix by pipetting. Transfer to a pre-chilled sterile tube.

9. Plate out the cells:

   (a) If using G418 selection, add 0.5 ml YPD containing 1 M sorbitol and incubate overnight at room temperature to allow expression of antibiotic resistance. Pellet the cells by centrifugation, resuspend in sterile water, and spread on to a single YPD + G418 agar plate.

   (b) If using auxotrophic selection, plate out directly on minimal selective agar (YNBD agar + supplements), ideally containing 1 M sorbitol (which increases efficiency ten-fold).

A much more rapid version of *Protocol 3*. which gives ten times lower efficiency, is to scrape a loop-full of cells off an agar plate, wash once in 1 M sorbitol by centrifugation, resuspend in 40 µl, and electroporate using up to 1 µg of DNA.

5. Techniques for induction and analysis of intracellular expression

As described above, it is generally desirable to use expression vectors with tightly regulated promoters such as *GAL* or *ADH2*, though constitutive promoters have been used successfully in many cases. Inductions can be carried out in minimal or rich media. However, growth rates and sometimes expression levels may be higher with the latter. The kinetics of promoter induction should be identical with different foreign genes, but the kinetics of accumulation of each protein will differ depending on its stability. Therefore an induction time course should be carried out to maximize yields. The procedures described in the following section are for 10 ml inductions but they can be scaled up without alteration. Induced cells may be washed with water and stored at $-70\,^{\circ}$C prior to analysis.

5.1 Galactose inducible promoters

GAL promoters are induced more than 1000-fold by addition of galactose but are strongly repressed by glucose, so that in glucose grown cultures maximal induction can be achieved only following depletion of glucose. Galactose inductions can be carried out in one of three ways:

(a) By growing the culture on a non-repressing carbon source (e.g. glycerol, lactate, or raffinose; raffinose gives the highest growth rates), then adding galactose, whereupon very rapid induction occurs (*Protocol 4A*). This method is the simplest and the most controllable and is recommended for most situations. It also the best method when short inductions are needed, e.g. with unstable proteins.

(b) By growing initially in glucose, then removing the glucose medium and changing to galactose medium (*Protocol 4B*). This leads to a lag of three to five hours in induction due to residual glucose repression. Growth in glucose minimizes expression before adding galactose, and should be used to prevent leaky expression of very toxic proteins.

(c) By growing cells in medium containing glucose and galactose, when the glucose is preferentially utilized before galactose induction can occur (*Protocol 4C*). This method is not very controllable but is simple and adequate for induction of stable proteins.

The first two methods are frequently used for small scale inductions but are impracticable for large scale inductions.

The methods given in *Protocol 4* are suitable for induction of G418 selectable vectors such as pWYG7L in YP-based media. When using auxotrophic selection markers, follow the same procedures but use YNB media containing the equivalent sugar. For *URA3* (e.g. pWYG4 or pBM150, *Figure 1*) or *TRP1* vectors use YNB + casamino acids. These protocols are also suitable for hybrid galactose inducible promoters (*Table 3*).

Protocol 4. Induction of *GAL* promoters

Equipment and reagents

- Sterile plastic 25 ml Universal containers (e.g. NUNC 3–64211)
- YP + G418, YPD + G418 (see *Table 5*)
- 20% galactose (<0.01% glucose, Sigma G0750; dissolve in water, autoclave or filter sterilize)
- 20% raffinose (dissolve in water, autoclave or filter sterilize)
- Sterile water

Protocol 4. *Continued*

A. *Induction from raffinose*

1. Prepare an overnight starter culture of the transformant in YP + 2% raffinose + G418. Measure the A_{600}.

2. Dilute the starter culture into 10 ml of YP + 2% raffinose + G418 in a 25 ml Universal container to an A_{600} of 0.25 (typically a 20-fold dilution). Incubate at 30 °C with shaking until the culture density reaches an A_{600} of approximately 1.0 (4–5 h).

3. Add 1 ml 20% galactose and incubate with shaking at 30 °C for 8–24 h (typically overnight). Measure the A_{600} before harvesting the cell pellet by centrifugation; the cultures should reach an $A_{600} = 10$–15.

B. *Induction from glucose*

1. Prepare a culture of the transformant in mid-exponential phase following steps 1 and 2 in part A, but in this case using medium containing glucose (i.e. YPD + G418).

2. Harvest the cells by centrifugation (20 °C, 2000 g, 5 min). Resuspend the pellet in 10 ml sterile water and re-centrifuge.

3. Resuspend the pellet in 10 ml YP + 2% galactose + G418 and incubate at 30 °C with shaking for 24–48 h. Measure the A_{600} before harvesting the cell pellet by centrifugation; the cultures should reach an $A_{600} = 5$–10.

C. *Induction of cultures grown on glucose and galactose*

1. Prepare an overnight starter culture of the transformant in YPD + G418. Measure the A_{600}.

2. Dilute the starter culture into 10 ml of YPD + 2% galactose + G418 in a 25 ml Universal container to an A_{600} of 0.25 (typically a 20-fold dilution).

3. Incubate at 30 °C with shaking for 24–48 h. Measure the A_{600} before harvesting the cell pellet by centrifugation; the cultures should reach an $A_{600} = 10$–15.

5.2 *ADH2* promoter

The *ADH2* promoter is repressed 100-fold by glucose. In order to maintain repression, cells must always be grown in excess glucose (e.g. 8%) since normal concentrations of glucose become depleted. It is also preferable to store transformants on plates containing 8% glucose. Inductions can be carried out in one of two ways:

(a) Cells are grown in medium containing excess glucose, then induced by changing to fresh medium containing a non-fermentable carbon source, e.g. ethanol, glycerol, raffinose (*Protocol 5A*). This method is the more well-controlled, and is suitable for carrying out reproducible time courses.

(b) Cells are grown initially in a lower concentration of glucose (e.g. 1%) which is gradually depleted leading to induction (*Protocol 5B*).

The methods in *Protocol 5* assume the use of a G418 selectable vector such as pWYG2L (*Figure 1*) but can be adapted for auxotrophic selection by replacing YP media by YNB.

Protocol 5. Induction of *ADH2* promoters

Equipment and reagents

- Sterile 25 ml Universal containers (e.g. NUNC 3–64211)
- Sterile 100 ml Erlenmeyer flask
- YP + G418 (see *Table 5*)
- 50% glucose (autoclave or filter sterilize)
- 20% raffinose (*Protocol 4*)
- Ethanol (AR)
- Diastix glucose test strips (Bayer Diagnostics 2804)

A. *Induction by changing to non-repressing carbon source*

1. Prepare a 10 ml exponential phase culture of the transformant in YP + 8% glucose + G418 (see *Protocol 4A*, steps 1 and 2).

2. Harvest the cells by centrifugation (20 °C, 2000 *g*, 5 min). Resuspend the pellet in 10 ml sterile water.

3. Harvest the cells by centrifugation and resuspend the pellet in 10 ml YP + 2% raffinose + 2% ethanol.

4. Incubate at 30 °C under conditions of maximum aeration (e.g. by incubating in a 100 ml Erlenmeyer flask with vigorous agitation) for 8–24 h. Induction starts after a lag of several hours. Measure the A_{600} before harvesting the cell pellet by centrifugation.

B. *Induction by depletion of glucose*

1. Prepare an overnight starter culture of the transformant in YP + 8% glucose + G418. Measure the A_{600}.

2. Dilute the starter culture into 10 ml of YP + 1% glucose + G418 to an A_{600} of 0.05 (typically a 100-fold dilution).

3. Incubate with vigorous aeration at 30 °C for 18–24 h. Glucose depletion can be monitored using Diastix glucose test strips. Measure the A_{600} before harvesting the cell pellet by centrifugation.

5.3 Glycolytic gene promoters

A number of commonly used expression vectors have promoters from glycolytic genes (e.g. *PGK*, *GAPDH*, *ADH1*); examples are the vectors pMA91 and pYcDE2 (*Figure 1*). These are most often used constitutively (see *Protocol 6*), although *PGK* expression vectors can be induced up to 20- to 30-fold by adding glucose to acetate grown cells (9). Constitutive high level expression using such vectors causes a metabolic burden on the cell that usually results in reduced growth rates. The degree of the effect will depend on the gene expressed, but doubling times may increase to three or four hours, or greater, in extreme cases. Very long doubling times are diagnostic of toxicity, and indicate that a regulated vector should be used.

Protocol 6. Constitutive expression using glycolytic gene promoters

Reagents

- YNBD + supplements (*Table 5*)

Method

1. Prepare an overnight starter culture of the transformant in YNBD + supplements. The doubling time will usually be longer than for the untransformed strain. Measure the A_{600}.

2. Dilute the starter culture into 10 ml of fresh medium to an A_{600} of 0.0025.

3. Incubate at 30 °C with shaking until the A_{600} reaches 0.25–0.5 (18–24 h depending on the foreign protein).

4. Harvest the cells by centrifugation (4 °C, 2000 g, 5 min) for analysis.

There is evidence that the use of higher glucose concentrations (e.g. 8%) can improve expression levels. Presumably this is because glucose then remains present throughout the growth/expression period.

5.4 Protein analysis

5.4.1 Disruption of yeast cells and preparation of protein extracts

Protein extracts may be made from yeast cells by:

- mechanical disruption
- sphaeroplasting followed by lysis in hypotonic buffer

The simplest method is to vortex the cells in the presence of glass beads (*Protocol 7*). This method can be used both in the analytical (e.g. 10 ml cultures) and preparative scale. Although proteases are less of a problem than with *E. coli*, cell breakage should be carried out in the cold, ideally in the presence of protease inhibitors.

The particular extraction buffer used will depend on the protein studied and the subsequent analysis or purification. Membrane proteins will be solubilized only in the presence of detergents, whereas nucleic acid binding proteins may be released only in high salt. If the aim is simply to detect a polypeptide in SDS–PAGE (either by staining or Western blotting) the composition of the buffer is less important, since the total extract will be denatured before analysis. However, the composition of the buffer may be more critical when analysis is by enzyme assay, ELISA, immunoprecipitation, or RIA.

Protocol 7. Preparation of protein extracts

Equipment and reagents

- 15 ml polypropylene snap-top tubes (Falcon 2059)
- Platform vortex mixer (e.g. IKA-Vibrax-VXR with test-tube rack, supplied by Merck-BDH 330/0360/00)
- Extraction buffer: 50 mM Tris–HCl pH 7.4, 100 mM NaCl, 2 mM EDTA, cooled on ice

- Glass beads (0.45 mm, B. Braun; leave 16 h in concentrated HCl, rinse thoroughly, and bake 16 h at > 150 °C)
- 5 × protease inhibitor cocktail: 20 mM EDTA, 20 mM EGTA, 20 mM PMSF, 10 mg/ml pepstatin, 10 mg/ml leupeptin, 10 mg/ml chymostatin, 10 mg/ml antipain; store at −20 °C

Method

1. Harvest the cells from a 10 ml induced culture by centrifugation (4 °C, 2000 g, 5 min). Resuspend the cell pellet in 10 ml water, transfer to a 15 ml snap-top tube, and centrifuge. Remove excess water from the pellet; the washed cell pellet can be stored at −70 °C.

2. Resuspend the cell pellet in 0.5 ml of ice-cold extraction buffer + 125 μl of 5 × protease inhibitor cocktail.

3. Add glass beads to two thirds the height of the meniscus, place the tubes on a platform vortex mixer in a cold room, and vortex at full speed for 10 min. [a]

4. Recover the cell extract from the glass beads using a fine-tipped plastic Pasteur pipette. To recover as much of the extract as possible, re-extract the beads with 0.5 ml buffer. Alternatively, centrifuge for 2 min at 1000 g through a plastic disposable filter (e.g. Poly-prep chromatography columns, Bio-Rad 731-1550; check that the protein of interest is not retained by the filter).

5. Determine the total protein concentration of the extract using a standard assay (e.g. Bio-Rad protein assay). The yield of total protein from a 10 ml culture grown to $A_{600} = 10$ is approximately 10 mg.

6. If desired, separate insoluble protein by centrifugation for 15 min at 4 °C in a microcentrifuge (12 000 g). [b] Transfer the soluble protein to a new tube and determine the protein concentration; keep this and the pellet containing insoluble proteins for analysis. [c]

Protocol 7. *Continued*

7. Store the protein extracts frozen at $-20\,^{\circ}$C or $-70\,^{\circ}$C. Upon thawing, a precipitate, largely consisting of cell membranes, may appear; this should be dispersed prior to SDS–PAGE.

---

[a] Samples can be vortexed by hand using a vortex mixer in six 30 sec bursts with 30 sec intervals on ice. Breakage can be checked by phase-contrast light microscopy, when intact cells appear bright, while broken cells are phase-dark 'ghosts'. Breakage is most efficient in wide round-bottom tubes. For larger volumes use 30 ml centrifuge tubes or a 'Bead Beater' (Biospec Products, PO Box 722, Bartlesville, Oklahoma 74005, USA).
[b] For large scale purification of proteins the preparation may be centrifuged at 100 000 *g* for 1 h to precipitate membranes.
[c] To determine the proportion of a protein that is soluble, resuspend the insoluble protein pellet in the same volume of buffer as the soluble protein preparation, and analyse equal volumes of 'total', 'soluble,' and 'insoluble' protein by SDS–PAGE.

5.4.2 Analysis of proteins

i. Analysis by SDS–PAGE

The simplest analysis is SDS–PAGE followed by staining of the gel with Coomassie blue. If the expression level is reasonably high ($>0.5\%$), or the protein migrates in a clear region of the gel, then it may be readily detected by comparison to proteins present in the induced host strain lacking the vector. For a clear band, quantitation may be achieved by densitometer scanning of the gel.

Where a suitable antibody is available, the protein may be detected in Western blots (see Chapter 4, *Protocol 6*). If the level is too low to quantitate on SDS–PAGE by staining, an estimate of the yield may be obtained by comparison to known amounts of the protein (if available) in Westerns. If no antibody is available, an alternative is to engineer the vector to produce a protein with an epitope tag that reacts with an available antibody. An estimate of the expression level may be obtained by comparison in Western blots to known amounts of another protein carrying the same epitope. Epitope tags and fusion proteins can also be used for rapid purification by affinity chromatography (ref. 29; also see Chapters 2 and 3).

ii. Immunoprecipitation

In order to increase the sensitivity of detection, cell proteins may be radiolabelled *in vivo* (see *Protocol 8*), concentrated by immunoprecipitation, and detected by autoradiography. Alternatively, in order to avoid *in vivo* labelling, the immunoprecipitated protein may be detected using a labelled or enzyme-conjugated antibody in Western blots (see Chapter 4, *Protocol 6*).

Immunoprecipitation of radiolabelled protein can be used to estimate the concentration of a specific protein (i.e. expression level), by measuring the proportion of immunoprecipitable to TCA precipitable radiolabel. Standard techniques may be used for immunoprecipitation of proteins extracted from yeast cells (ref. 30; see also Chapter 4, *Protocol 7*).

Protocol 8. *In vivo* labelling of yeast proteins

Reagents

- YNB (see *Table 5*) containing inducer plus all of the supplements listed except L-methionine, and the relevant supplement for auxotrophic selection of the expression vector

- L-[^{35}S]methionine (e.g. Amersham *in vivo* cell labelling grade, SJ1015)
- Sterile water

Method

1. Carry out an induction of a 10 ml culture as described in previous protocols (the method is best suited to *Protocols 4A* or *6*); rich medium can be used at this stage. After 2 h, while the cells are still in the mid-exponential phase, harvest them by centrifugation (20 °C, 2000 g, 5 min).

2. Resuspend the cell pellet in 10 ml sterile water. Harvest by centrifugation and resuspend the pellet in 10 ml YNB + all the necessary supplements + inducer. Incubate at 30 °C with shaking for 20 min.

3. Add 10–50 µCi [^{35}S]methionine/ml (i.e. 100–500 µCi) and incubate for a further 2 h.

4. Harvest the cells by centrifugation. Resuspend the cell pellet in 10 ml sterile water and recentrifuge.

5. Use the washed cell pellet for preparation of a protein extract (*Protocol 7*). For secreted proteins, also save the labelling medium for direct analysis.

6. To determine the *in vivo* half-life of a protein, carry out step 3 then:

   (a) Add an excess of unlabelled methionine (e.g. 20 mg/litre) to the culture.

   (b) Harvest cells at several different time points (e.g. 0.5–12 h), and prepare protein extracts.

   (c) Quantitate the amount of the protein in question at these time points. This is most readily carried out by immunoprecipitation of equal amounts of cell protein, followed by SDS–PAGE and auto-radiography. The amount of undergraded protein can be determined by densitometry of the autoradiograph, and the rate of disappearance of the protein can be used to calculate its half-life.

---

6. Preparation and analysis of secreted proteins

6.1 Detection of secreted proteins

The techniques for inducing expression of secreted proteins are the same as those described above. Efficiently secreted proteins may accumulate to

5–50 μg/ml at normal laboratory culture densities, so that they account for most of the secreted yeast protein, and this concentration may increase proportionately in high density fermentations (up to 30-fold). In such cases the protein can be analysed directly in the culture supernatant, especially if a specific assay is available such as RIA, ELISA (see Chapter 4), or enzyme activity. However, for the initial analysis of a new secreted protein by SDS–PAGE it is advisable to concentrate the medium. This is most conveniently achieved by ultrafiltration, using membranes that retain molecules above a certain molecular weight (*Protocol 9A*). Alternatively secreted proteins may be concentrated by TCA precipitation (*Protocol 9B*). In order to achieve still greater concentration, by simulating high density fermenter cultures, cells may be concentrated 20-fold prior to induction.

Protocol 9. Preparation of secreted proteins

Equipment and reagents

- Centricon 3, 10, 30, or 100 (from Amicon), depending on the M_r of the foreign protein
- Induced culture (see *Protocol 4, 5, or 6*)
- TCA/deoxycholate: 50% trichloroacetic acid (TCA) containing 0.2% sodium deoxycholate
- 10% TCA
- Acetone
- 1 × SDS sample buffer: 50 mM Tris–HCl pH 6.8, 100 mM DTT, 2% SDS, 0.1% bromophenol blue, 10% glycerol

A. *Concentration of secreted proteins by ultrafiltration*

1. Induce the secretion vector using a standard induction protocol (*Protocols 4, 5*, or *6*).[a,b]

2. Pellet the cells by centrifugation (4 °C, 2000 *g*, 5 min). Retain the supernatant (culture medium) for analysis of secreted product.[c]

3. Concentrate the medium using Centricon tubes with an appropriate molecular weight cut-off (e.g. Centricon 30 for proteins of > 30 kd), according to the manufacturer's instructions. An angle rotor must always be used to prevent complete dehydration of the sample. Since the length of centrifugation required may vary, centrifuge for a shorter period at first, and then repeat if necessary. The following are guidelines for Centricon 30: using YP media, 2.5 ml of medium can be concentrated to 250 μl by centrifuging for 40 min at 5000 *g*; with YNB media, 2.5 ml can be concentrated to 50 μl in 25 min.

B. *Concentration of secreted proteins by TCA precipitation*

1. Add 0.25 vol. of TCA/deoxycholate to the culture supernatant prepared as in part A, and leave on ice for 30 min.

2. Collect the precipitate by centrifugation (4 °C, 20 000 *g*, 20 min).

3. Resuspend the precipitate in 10% TCA then centrifuge. Repeat this process using acetone. Drain the pellet and allow to dry.

4. Re-dissolve the protein pellet in 1 × SDS sample buffer ready for SDS–PAGE analysis. Protein from up to 10 ml of culture may be applied to one SDS–PAGE sample well.

[a] Several secreted proteins have been found to be accumulate to lower levels in minimal medium than in rich medium. Yields in minimal medium have been improved by adding casamino acids, or buffering to pH 6.0 using potassium phosphate (unbuffered media reach pH ≈ 4).

[b] In order to concentrate the secreted proteins further, concentrate the cells 20-fold by centrifugation prior to induction.

[c] The cell pellet should also be stored to allow analysis of protein that has accumulated intracellularly.

Since yeast cells secrete few proteins, a foreign protein is readily identified by comparing proteins in the culture supernatant of transformants and untransformed strains by SDS–PAGE. Faint bands may be detected by silver staining or Western blotting.

Different foreign proteins are secreted into the culture medium with different efficiencies. In most cases some of the product will have accumulated, or will be in transit, within the cell (or cell wall). The proportion in the medium ('secretion efficiency') may be determined by preparing a protein extract from the cell pellet (see *Protocol 7*) and comparing the amount of product in this extract to that secreted in an equivalent culture volume. If this analysis is carried out by a method other than SDS–PAGE, then the intracellular protein must be released from membrane-bound secretory organelles using non-ionic detergent (e.g. 0.5% Triton X-100) added to the extraction buffer (*Protocol 7*).

6.2 Analysis of glycoproteins

Secreted foreign proteins may be glycosylated either because they contain glycosylation sites that are used naturally, or because they have sequences fortuitously recognized as sites for *N*- or *O*-linked glycosylation in yeast. The signal for *N*-linked glycosylation in yeast (Asn–X–Ser/Thr) is the same as for other eukaryotes. Therefore fortuitous *N*-linked glycosylation is usually only found on proteins that are not naturally secreted, or on bacterial secreted proteins. *O*-Linked glycosylation is less common than *N*-linked, and though the signals for this are poorly defined, there are clearly differences between yeast and higher eukaryotes. Differences in the degree glycosylation between the yeast-derived and native product may be identified, in the first instance, by comparison of their molecular weight by SDS-PAGE (e.g. tetanus toxin fragment C, *Figure 6*).

N-Linked glycosylation in *S. cerevisiae* is of two types: high mannose (8–14 mannose residues) or mannan (containing an outer mannose chain of up to 75 residues). The presence of mannan on foreign glycoproteins ('hyperglycosylation') is undesirable since it may mask surface features of the protein. Hyper-

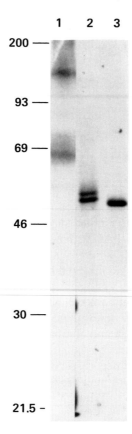

Figure 6. Coomassie blue stained SDS–PAGE of secreted tetanus toxin fragment C from *S. cerevisiae* (31). Lane 1—concentrated culture supernatant showing both glycosylated and hyperglycosylated forms of fragment C. Lane 2—same as lane 1 but deglycosylated with Endo H (see *Protocol 10*). *N*-terminal sequence analysis indicated that the upper band of the doublet is due to the presence of *N*-terminal Glu–Ala pairs, derived from incompletely processed α-factor leader. Lane 3—native (unglycosylated) fragment C for comparison.

glycosylated proteins migrate in SDS–PAGE as diffuse bands of very high molecular weight (e.g. fragment C, *Figure 6*). They frequently react poorly with antibodies in Western blots and may be difficult to detect by staining. Therefore secreted glycoproteins should also be analysed after deglycosylation, which should generate discrete polypeptide species, using an enzyme such as endoglycosidase H (Endo H), which removes *N*-linked carbohydrate. SDS–PAGE of partially deglycosylated glycoproteins can result in a ladder of bands corresponding to the number of glycosylation sites in a molecule. Intracellularly accumulated glycoprotein may also be analysed using endoglycosidase H, by preparing cellular protein extracts and deglycosylating as in *Protocol 10*.

Protocol 10. Deglycosylation of glycoproteins using endoglycosidase H

Reagents

- Endoglycosidase H (Boehringer-Mannheim)
- Endo H buffer: 200 mM NaH$_2$PO$_4$ pH 5.5, 10 mM 2-mercaptoethanol, 1% SDS
- 5 × protease inhibitor cocktail (*Protocol 7*)

Method

1. To 25 μl of sample (up to 100 μg of protein) add 5 μl of Endo H buffer, and boil for 5 min.[a]

2. Cool the sample on ice, and then add 10 μl of 5 × protease inhibitor cocktail and 10 μl (10 mU) Endo H. Incubate at 37 °C overnight.

3. Compare the native and deglycosylated proteins by SDS–PAGE.[b]

[a] To determine whether carbohydrate residues are external or internal (cryptic), leave out the SDS and 2-mercaptoethanol and do not boil; only external carbohydrate will be removed. This can be used to prepare undenatured deglycosylated protein.
[b] Endo H leaves a single *N*-acetylglucosamine residue at each site, so that the deglycosylated polypeptide will have a slightly greater molecular weight than predicted.

Mutant strains may be used in order to produce glycoproteins that are not hyperglycosylated (*mnn9*; see *Table 4*), though these strains are not generally robust. Glycoproteins secreted from *mnn9* strains have no outer chain mannose residues, but are still antigenic due to the presence of terminal α1,3 mannose linkages. The *mnn1 mnn9* strain, which produces proteins lacking these residues, can be used to produce non-antigenic, non-hyperglycosylated glycoproteins. Glycoproteins produced in the *pmr1* strain (see Section 7.7) are also non-hyperglycosylated.

7. Further analysis of expression, diagnosis of problems, and optimization

7.1 Introduction

Insertion of a foreign gene into an expression vector does not guarantee a high yield of the protein; problems can arise at numerous stages from transcription of the gene through to protein turnover. The following problems are among the most common:

- inability to produce some proteins because of their high toxicity
- failure to express AT-rich genes due to premature transcriptional termination
- inefficient secretion of some larger (> 30 kd) proteins
- inappropriate glycosylation of mammalian glycoproteins

The separate stages involved in foreign gene expression are discussed below, along with the particular problems that may occur at each stage, techniques for diagnosing them, and potential solutions.

7.2 Genetic stability of the vector

Low yields of foreign proteins may arise because of reduced copy number, re-arrangement, or mutation of the vector (1). This most frequently arises from the constitutive expression of toxic proteins, which results in reduced growth rate and selection for dramatically reduced vector copy number (e.g. from 50 to 1 copy/cell). Forced high copy maintenance (e.g. with *LEU2*-d) prevents this but may then cause selection for FLP-mediated structural rearrangement of the vector, eliminating foreign gene expression. Vectors derived from pJDB219 (e.g. pWYG7L) can recombine with native 2μ plasmid resulting in the elimination of non-yeast sequences but retention of the *LEU2*-d marker, so that leucine selection is ineffective. However, this problem can be eliminated by using vectors where the selection marker lies outside 2μ sequences (e.g. pWYG7L using G418 selection), so that the foreign DNA and selection marker remain linked. The use of vectors in $2\mu^0$ strains eliminates many of these problems. However, other structural rearrangements or even point mutations in the foreign gene may then be seen, if the selection pressure is great.

Genetic changes may also occur in the host cell, for example reversion of the auxotrophic mutation or integration of the plasmid selection marker, resulting in plasmid loss. These events may be prevented by using a host strain with a deletion of the auxotrophic marker locus, or a recombination-deficient strain (1). There is also an example of a host mutation resulting in inactivation of the promoter of the expression vector (1). Most of the changes described can be avoided by using a tightly regulated promoter.

7.2.1 Techniques for DNA analysis of transformants

DNA analysis of transformants may be used to verify the integrity or deter-mine the copy number of episomal vectors. It is also carried out routinely to verify the structure of chromosomal integrants. *Protocol 11* describes the preparation of DNA from yeast for such analysis.

Protocol 11. Preparation of DNA from yeast for Southern blot analysis

Equipment and reagents

- Glass Pasteur pipettes
- SCE buffer: 1 M sorbitol, 10 mM sodium citrate, 10 mM EDTA
- 1 M DTT (stored at −20 °C)
- Zymolyase-100T, 3 mg/ml (ICN 32–093–1, make up fresh)
- 1% SDS
- 5 M potassium acetate
- TE buffer: 10 mM Tris–HCl pH 8.0, 0.1 mM EDTA
- Ribonuclease A, 10 mg/ml, DNase-free (dis-solve in TE buffer containing 15 mM NaCl, boil 10 min, store at −20 °C)
- Isopropanol (2-propanol)

Method

1. Prepare a 40 ml overnight culture of the transformant grown in selective medium (stable transformants can be grown in YPD).

2. Harvest the cells by centrifugation (20 °C, 2000 g, 5 min). Resuspend the pellet in 40 ml SCE buffer and re-centrifuge.

3. Resuspend the pellet in 2 ml SCE buffer + 20 μl 1 M DTT.

4. Add 100 μl of zymolyase and incubate at 37 °C for 1 h or until sphaeroplast formation approaches 100%. (The efficiency of sphaeroplast formation can be monitored using phase-contrast microscopy. Place 10 μl on a microscope slide, mix with 10 μl 1% SDS. Intact cells appear phase-bright while lysed 'ghosts' are phase-dark.)

5. Add 2 ml of 1% SDS to the sphaeroplasts, mix gently by inversion, and leave on ice for 5 min.

6. Add 1.5 ml of 5 M potassium acetate, mix by inversion, and centrifuge at 10 000 g for 10 min.

7. Add 2 vol. of ethanol to the supernatant, leave at room temperature for 15 min, then centrifuge at 10 000 g for 15 min.

8. Drain the pellet and allow to dry. Add 3 ml of TE buffer and 150 μl 10 mg/ml ribonuclease A. Shake gently at 37 °C for 1–2 h to resuspend the pellet and digest RNA. If the sample still contains any insoluble material it can be removed by centrifuging at 10 000 g for 15 min.

9. Slowly add an equal volume of 2-propanol to the DNA solution so that it forms a separate layer. Mix the layers by shaking gently: a 'spool' of DNA should form.

10. Remove the spool using a glass hook (formed from a Pasteur pipette in a Bunsen flame), drain off excess liquid, and drop the spool into 250–500 μl TE buffer.[a] The spooled DNA should dissolve readily; 10 μl of the solution should be ample for digestion and analysis by Southern blotting.

[a] Spooling gives the highest quality of DNA for restriction analysis. If the DNA precipitate is too loose to be spooled, it can be concentrated by centrifugation and spooled from a smaller volume of TE buffer. Alternatively, if spooling proves difficult, the DNA can be ethanol precipitated in the usual way; in this case it will require phenol extraction, and the final ethanol precipitate may need to be dissolved in TE buffer by shaking overnight at 4 °C.

The DNA is digested using suitable restriction enzymes and analysed by Southern blotting to determine the copy number and site of integration of integrating vectors, or to confirm that gene replacement has occurred with transplacement vectors. The copy number of episomal vectors can also be estimated in Southern blots. This is done in the following way:

(a) Digest total DNA with a restriction enzyme to generate fragments containing the selection marker (e.g. *URA3*). These should have a different size depending on whether they originated from the plasmid or chromosomal DNA.

(b) Separate the digest by electrophoresis, Southern blot the DNA fragments, and probe with radiolabelled DNA which hybridizes to the selection marker.

(c) Measure the relative intensities of the two bands (plasmid/chromosomal) to determine the plasmid copy number. The most accurate quantitation is by phosphorimaging of the Southern blot, or scintillation counting of excised bands. Quantitation by densitometry of autoradiographs introduces errors at high copy numbers. Errors due to a difference in Southern transfer efficiency are also introduced when the two fragments have very different sizes.

In order to confirm the integrity of an episomal vector in a transformant, the plasmid may be extracted and rescued by transformation of *E. coli* (e.g. to kanamycin resistance if using pWYG7L). A number of *E. coli* transformants may then be analysed (e.g. using restriction digests, or by sequence analysis) to determine whether rearrangements had occurred in yeast cells. The rapid DNA preparation given in *Protocol 12* is ideal for this (it can also be used for Southern blot analysis instead of DNA prepared as in *Protocol 11*).

Protocol 12. Rapid preparation of DNA from yeast

Equipment and reagents

- Glass beads (*Protocol 7*)
- Extraction buffer: 10 mM Tris–HCl pH 8.0, 1 mM EDTA, 100 mM NaCl, 0.1% SDS
- TE buffer: 10 mM Tris–HCl pH 8.0, 1 mM EDTA
- Phenol/chloroform (a 25:24:1 mixture of phenol equilibrated with Tris–HCl pH 8.0, chloroform, and isoamyl alcohol)

Method

1. Scrape a colony of a transformant from an agar plate and resuspend in 200 μl of extraction buffer in a 1.5 ml microcentrifuge tube.

2. Add glass beads to just below the level of liquid and vortex vigorously for 1 min.

3. Add 200 μl of phenol/chloroform, vortex briefly, separate the phases by centrifugation, and remove the aqueous layer.

4. Re-extract the aqueous layer with phenol/chloroform, then add 2 vol. of ethanol, and store at −70 °C for 20 min to precipitate nucleic acid.

5. Collect the precipitate by centrifugation, remove excess liquid, and resuspend the pellet in 20 μl of TE buffer.

Note that in cells expressing FLP recombinase, 2μ vectors will exist in two forms (A and B, see *Figure 2*). In strains containing both a 2μ vector and native 2μ, co-integrants will be seen. These rearrangements are important only if they result in loss of the foreign gene.

7.3 mRNA synthesis

i. Transcriptional initiation

It is generally assumed that the steady-state mRNA level is a primary determinant of the final yield of a foreign protein, and that the most important factor is the rate of transcriptional initiation. Results with over 20 genes expressed using multicopy *PGK* promoter vectors have shown that foreign mRNA levels are usually approximately tenfold lower than *PGK* mRNA, possibly due to the presence of enhancer elements within the *PGK* coding sequence. Therefore, transcriptional initiation may frequently limit expression of foreign genes in *S. cerevisiae*, so that the very high yields obtained in *E. coli* or baculovirus are rarely possible. This transcriptional effect does not appear to occur using the *AOX1* promoter in the yeast *P. pastoris*, and indeed much higher expression levels are often seen, with multicopy transformants, in this yeast (see later, *Table 6*).

Using galactose regulated promoters, promoter activity becomes severely limited by GAL4 *trans*-activator concentration at multiple copy numbers. This limitation of galactose regulated promoters is usually tolerated. However, it may be alleviated by over-expression of *GAL4*, though tight regulation is then lost (1); *GAL* expression vectors containing *GAL4* are available for this purpose (5). Regulated over-expression of *GAL4* has been used to increase promoter activity and maintain tight regulation, though we have found such systems to behave irreproducibly (1).

ii. Premature termination of transcripts

The production of full-length transcripts can be affected by the presence of fortuitous sequences in the foreign gene that cause termination or polyadenylation. Though not widely recognized until recently, there is evidence that this is a widespread problem in yeast that may result in complete failure or lowered efficiency of gene expression (1). Genes having an AT content in excess of 65% usually contain multiple terminators/polyadenylation sites so that full-length mRNA will not be produced (e.g. fragment C, *Figure 7*). However, we have also seen the problem with a number of genes of average AT content that have AT-rich tracts.

Termination/polyadenylation signals of yeast genes are found in AT-rich DNA, but they are poorly defined, so that it is not possible to predict the location of fortuitous signals in AT-rich foreign genes. Therefore at present the only reliable means of solving this problem is to chemically synthesize the gene, or offending portions, using synonymous codons so as to increase the

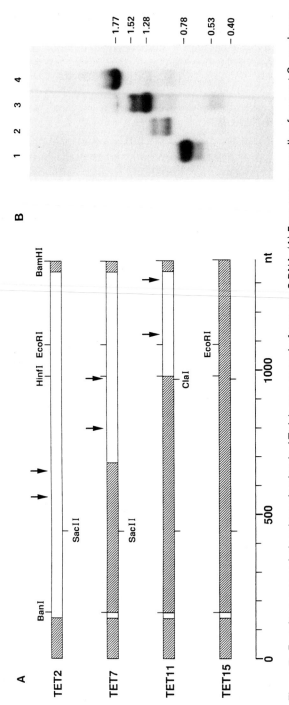

Figure 7. Fortuitous transcriptional termination in AT-rich tetanus toxin fragment C DNA. (A) Four genes encoding fragment C are shown, with synthetic DNA of increasing GC content (hatched) achieved using GC-rich synonymous codons. The approximate 3' ends of transcripts generated in yeast are indicated. (B) Northern blot showing fragment C-specific RNA from TET2, TET7, TET11, and TET15 (lanes 1–4, respectively): only TET15 gave abundant full-length mRNA.

GC content (*Figure 7*; ref. 31). Genes synthesized using the subset of codons found in highly expressed yeast genes (32) should be free of termination/polyadenylation signals provided there are no runs of As or Ts. Before embarking on the chemical synthesis of a gene, it is advisable to attempt to define the problem region as accurately as possible by transcript mapping.

iii. mRNA instability

There are few examples of low mRNA stability being a major factor affecting product yield. However, in one example the presence of a foreign 3' UTR resulted in low mRNA levels and low protein yield (33). It is therefore prudent to eliminate 3' untranslated DNA.

7.3.1 Techniques for RNA analysis

The mRNA should be examined if no protein product is detected. The simplest method is to prepare total RNA and analyse this by Northern blotting. With constitutive expression systems, RNA should be prepared from exponentially growing cells, when the proportion of mRNA to total RNA is highest. When using the rapidly induced *GAL* system, RNA may be prepared either one to four hours after induction, or after a longer period (e.g. overnight incubation), provided that the cells are still growing exponentially. Cells can be washed and stored at $-70\,^{\circ}\text{C}$ prior to the RNA preparation. The method given in *Protocol 13* reliably gives intact RNA, but the usual precautions should be taken in the subsequent handling of the RNA (i.e. wear gloves, treat buffers and tubes with diethylpyrocarbonate (DEPC) to destroy ribonuclease).

Protocol 13. RNA preparation

Equipment and reagents

- 15 ml polypropylene tubes with caps (Sarstedt 55.510, soaked overnight in 0.1% DEPC then autoclaved)
- 1.5 ml microcentrifuge tubes (soaked overnight in 0.1% DEPC then autoclaved)
- Platform vortex mixer (e.g. IKA-Vibrax-VXR with test-tube rack, supplied by Merck-BDH 330/0360/00)
- Glass beads (*Protocol 7*)
- DEPC-treated water (add 0.1% DEPC to distilled water, mix by shaking; loosen the lid, leave overnight, then autoclave)

- Guanidinium thiocyanate solution (dissolve 50 g guanidinium thiocyanate, 0.5 g sodium N-laurylsarcosinate, in 80 ml water. Add 2.5 ml 1 M sodium citrate pH 7.0, 0.7 ml 2-mercaptoethanol. Make up to 100 ml with water and filter. Store at 4°C).
- Phenol/chloroform (*Protocol 12*)
- Chloroform/isoamyl alcohol (24:1)
- 1 M acetic acid (made in DEPC-treated water)
- 3 M sodium acetate (pH 4.5 by adding acetic acid)

Method

1. Prepare a 50 ml galactose induced yeast culture. Either induce as in *Protocol 4C* but harvest cells after 2–4 h, or induce for 16 h as in

---

Protocol 13. *Continued*

Protocol 4C but alter the culture density, so that the final density has $A_{600} < 2$. For constitutive systems prepare a 50 ml culture of $A_{600} < 2$.

2. Harvest the cells by centrifugation (4 °C, 2000 *g*, 10 min). Resuspend the pellet in 10 ml ice-cold water and re-centrifuge.

3. Resuspend the pellet in 3 ml guanidinium thiocyanate solution and transfer the suspension to a DEPC-treated 15 ml Sarstedt tube.

4. Add glass beads to two thirds the height of the meniscus, place the tubes in a platform vortex mixer in a cold room, and vortex at full speed for 15 min.

5. Add 1 vol. of phenol/chloroform to the lysate, vortex again, and centrifuge (10 000 *g*, 15 min). Keep the upper phase and transfer to a new Sarstedt tube.

6. Re-extract the aqueous phase repeatedly with phenol/chloroform as in step 5, until there is little or no material at the interface.

7. Extract once with 1 vol. of chloroform/isoamyl alcohol.

8. Transfer the aqueous phase to a clean DEPC-treated tube and precipitate RNA by adding 0.025 vol. of 1 M acetic acid and 0.75 vol. of ethanol. Leave at −20 °C at least 30 min.

9. Collect the precipitate by centrifugation (10 000 *g*, 15 min) and drain off excess liquid. Vortex the pellet in 70% ethanol (in DEPC-treated water), centrifuge, and remove excess liquid.

10. Dissolve the precipitate in 0.4 ml DEPC-treated water. Transfer the solution to a microcentrifuge tube and re-extract once with phenol/chloroform, then once with chloroform/isoamyl alcohol. Precipitate by adding 0.1 vol. of 3 M sodium acetate and 2.5 vol. of ethanol.

11. Collect the precipitate by centrifugation in a microcentrifuge (12 000 *g*) for 15 min.

12. Remove excess liquid and resuspend the RNA in 100 μl DEPC-treated water. Store at −70 °C. The original 50 ml culture (at $A_{600} = 1$) should yield 150–300 μg of RNA; 5–20 μg per track should be used for gel electrophoresis and Northern blotting.

Analysis of the RNA by agarose gel electrophoresis and Northern blotting is usually sufficient to determine whether the expected full-length transcript is produced. Transcripts can be mapped by using small fragments, derived from the 5′ and 3′ UTR (i.e. vector promoter or terminator DNA), or from the cloned gene, as hybridization probes.

7.4 Polypeptide synthesis

Translational efficiency is mainly controlled by the rate of initiation, and in yeast this is greatly inhibited by upstream AUGs, runs of G, or secondary structure in the 5′ untranslated leader of an mRNA. For example, upstream viral sequences inhibited overall expression of hepatitis B core antigen by 500-fold (34). Translation of foreign transcripts should present no problems provided non-vector sequences 5′ to the gene are eliminated (see Section 2.5). There is some evidence that the presence of rare codons may limit yields in some situations, so that it may be worthwhile chemically synthesizing small genes with codons found in highly expressed yeast genes (32), in the hope of increasing yield.

7.5 Protein folding, processing, and stability

i. Protein folding

Most foreign cytoplasmic proteins are correctly folded in yeast, and indeed an important use of yeast expression systems is in producing proteins that accumulate as insoluble aggregates in *E. coli*. In contrast, normally secreted proteins encounter an abnormal environment in the cytoplasm and often become insoluble (e.g. HSA), though many are soluble and active (e.g. α-interferon). Where there is a tendency to aggregation, the proportion of soluble product might be improved by: reducing growth rate, reducing temperature, reducing rate of induction.

ii. Post-translational processing

An advantage of yeast in expressing mammalian proteins is in its ability to carry out similar post-translational modifications, for example N-terminal acetylation, phosphorylation, myristylation, and isoprenylation (1). Removal of N-terminal methionine and/or acetylation are predictable events that depend on the penultimate amino acid residue (1).

iii. Protein stability

The few examples of very high level expression (> 25%) in *S. cerevisiae* are of unusually stable proteins, suggesting that protein half-life is a factor of major importance. Very low yields are obtained with proteins which are naturally short-lived (e.g. regulatory proteins) or with some polypeptides that are naturally secreted. Possible solutions are:

- fusion of the protein to a stable carrier such as superoxide dismutase or glutathione S-transferase (1)
- induction at a lower temperature
- use of protease-deficient host strains
- use of very short induction times.

7.5.1 Techniques for measuring protein stability

The stability of a protein can be readily measured by pulse-chase radiolabelling experiments, if antibodies are available for immunoprecipitation (see *Protocol 8*). Alternatively, the stability of a protein expressed using the *GAL* or *ADH2* promoters can be estimated by adding glucose to induced cultures in order to rapidly shut off expression. The decay rate may then be followed by quantitation of the protein relative to total protein, using any appropriate detection method (e.g. SDS–PAGE followed by staining or Western blotting), at different times.

7.6 Processing of secreted proteins

Secretion of proteins involves a complex series of events involving several membrane-bound organelles. Following co-translational insertion into the endoplasmic reticulum, proteins are processed in number of ways including: signal peptide cleavage, polypeptide folding, disulfide bond formation, glycosylation, and, in some cases, proteolytic processing. Therefore, in addition to confirming that the approximate size of a secreted product is correct, it is often important to confirm the authenticity of the N- and C-termini, disulfide bonding, etc. N-terminal sequence analysis can be readily carried out on polypeptides separated by SDS–PAGE (e.g. fragment C, *Figure 6*). Small polypeptides can be purified by HPLC before N-terminal sequence analysis and amino acid analysis (e.g. EGF, *Figure 8*). The following types of abnormal processing may occur:

(a) Incorrect cleavage of a preprotein resulting in a product with additional or missing N-terminal amino acids. This is more likely to occur with non-yeast signal peptides.

(b) Incomplete removal of the Glu–Ala spacers in α-factor preproteins, so that a proportion of the secreted product has additional N-terminal Glu–Ala pairs. This situation is very common (e.g. with fragment C, *Figure 6*). It can be solved by constructing a vector not encoding the Glu–Ala spacer, or possibly by over-expressing *STE13*.

(c) Incomplete KEX2 processing of α-factor fusions, so that the preproprotein (usually in a hyperglycosylated form) is secreted. This problem may arise when the Glu–Ala spacers are deleted. It may be solved by over-expression of *KEX2* or *YAP3* protease (1).

(d) Aberrant endoproteolysis of the foreign protein, by *KEX2* or other proteases, to generate fragmented product (1). These problems may be alleviated by trying different host strains.

(e) C-terminal trimming resulting from carboxypeptidase activity (e.g. with EGF, *Figure 8*). This problem may be eliminated through the use of carboxypeptidase mutant strains (1).

7.7 Strategies for improving secretion efficiency

Inefficient secretion is a common problem with large heterologous proteins. In attempting to maximize levels of secretion, significant improvements can often be made using the following empirical approaches:

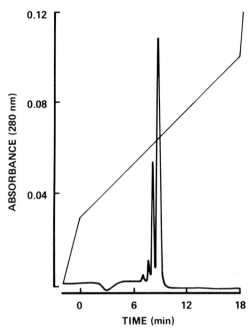

Figure 8. HPLC analysis of murine EGF secreted by *S. cerevisiae* (35). YNB culture supernatant was subjected to C18 reverse-phase chromatography, and eluted in a 10–50% gradient of acetonitrile + 0.1% trifluoroacetic acid. *N*-terminal sequence analysis and amino acid analysis were performed on the fractions. The results suggested that the four peaks correspond to (from *right* to *left*): (1) EGF lacking C-terminal Arg[53] (EGF 1–52; due to KEX1 processing), (2) EGF lacking Leu[52] and Arg[53] (EGF 1–51), (3) EGF 1–52 with oxidized Met[21], (4) EGF 1–51 with oxidized Met[21].

(a) Testing different promoter strengths. It has been reported that strong promoters may reduce yields of secreted proteins by causing toxicity and reduced plasmid copy number. Promoters of moderate strength such as *MFα1* may give higher yields.

(b) Testing different signal peptides. Several-fold differences in yield have been seen in some cases (e.g. HSA; 36).

(c) Testing many different strains, since they may show a very wide variation in yield.

(d) Optimizing medium and physiological conditions. Some proteins only accumulate efficiently in rich media or minimal media buffered to near neutral pH (e.g. pH 6). Some proteins may be more efficiently secreted at lower temperatures (e.g. 25 °C).

(e) Use 'super-secretion' mutants, e.g. *pmr1* strains (*Table 4*), or, if feasible, select for 'super-secretion' of the protein in question; for some secreted proteins, dramatic increases may be made. The general strategy of repeated rounds of mutagenesis followed by screening colonies for increased secretion has often been successful (e.g. with HSA; 37), but this depends on having a simple colony screen for the secreted product.

(f) Try alternative yeast species, e.g. *P. pastoris* or *Kluyveromyces lactis*, since there are examples of proteins that are secreted much more efficiently in these yeasts (1).

7.8 Toxicity

High level expression of a foreign gene places a metabolic burden on the host cell, reducing its growth rate and the efficiency of gene expression. Expression of some genes causes a more acute effect, either through a severe effect on metabolism or by direct toxicity. A consequence of this is the inadvertent selection of variants expressing lower levels of protein, particularly when constitutive expression systems are used. Even with regulated promoters, gradual accumulation of a toxic protein through leaky expression may have the same effect (with *GAL* promoters this can be minimized by having 8% glucose in propagation media).

Toxicity is common with membrane proteins or complex secreted proteins since these may accumulate as malfolded aggregates, blocking the secretory pathway. The over-expression of transcriptional *trans*-activators may also result in toxicity, through their ability to sequester transcriptional factors. The following approaches have been successful in overcoming the problem in some cases:

● secretion of the product, e.g. with HIV protease (21)
● selection of resistant mutations in the host strain, e.g. with insulin-like growth factor-1 (38).

However, toxicity can be an intractable problem, and it may then be more profitable to try a transient expression system, such as baculovirus.

8. Concluding remarks

In the early stages, the use of yeast for the expression of foreign genes was very attractive for the reasons mentioned in the introduction to this chapter. However, some of the disadvantages have become more apparent with the advent of powerful alternative systems, such as baculovirus. The most significant disadvantage, the substantially lower percentage yields, has now been addressed by the use of novel, regulated promoter systems for *S. cerevisiae* (*Table 3*), and by the use of yeasts such as *P. pastoris* which naturally have powerful, tightly regulated promoters. *Table 6* shows a comparison of the expression of two proteins in four systems, *S. cerevisiae*, *P. pastoris*, *E. coli*, and baculovirus, illustrating the comparative strength of the *P. pastoris* multicopy *AOX1* system. Alternative yeasts such as *P. pastoris, K. lactis,* and

Table 6. Comparison of the expression of two genes in
E. coli, Baculovirus, *S. cerevisiae*, and *P. pastoris*

	Expression level	
Expression system	Fragment C	Pertactin
E. coli	24%	30%
Baculovirus	10%[a]	>40%[a]
S. cerevisiae GAL7	2–3%	0.1%
P. pastoris	28%[b]	10%[c]

[a] Estimate; yield variable and greatly reduced on scale up.
[b] Scale up without loss of yield to > 12 g/litre.
[c] Scale up without loss of yield to > 4 g/litre.

Hansenula polymorpha are finding increasing use, especially for industrial production of proteins. However, *S. cerevisiae* remains the simplest and most versatile system for general use because of the huge range of genetic markers, vectors, and techniques that are available. In this chapter we have attempted to list the basic approaches and techniques required for the expression of cloned genes in yeast. Other texts should be referred to for more details on yeast genetics and manipulation of yeast chromosomes (39), and for further information on foreign gene expression (1, 40).

References

1. Romanos, M. A., Scorer, C. A., and Clare, J. J. (1992). *Yeast*, **8**, 423.
2. Hadfield, C., Cashmore, A., and Meacock, P. A. (1986). *Gene*, **45**, 149.
3. Kingsman, S. M., Cousens, D., Stanway, C. A., Chambers, A., Wilson, M., and Kingsman, A. J. (1990). In *Methods in enzymology* (ed. D. V. Goeddel), Vol. 185, pp. 329–41. Academic Press, London.
4. Johnston, M. and Davis, R. W. (1984). *Mol. Cell. Biol.*, **4**, 1440.
5. Rose, A. B. and Broach, J. R. (1990). In *Methods in enzymology* (ed. D. V. Goeddel), Vol. 185, pp. 234–79. Academic Press, London.
6. Lopes, T. S., Klootwijk, J., Veenstra, A. E., van der Aar, P. C., van Heerikhuizen, H., Raue, H. A., and Planta, R. J. (1989). *Gene*, **79**, 199.
7. Bijvoet, J. F. M., van der Zanden, A. L., Goosen, N., Brouwer, J., and van de Putte, P. (1991). *Yeast*, **7**, 347.
8. Rosenberg, S., Coit, D., and Tekamp-Olson, P. (1990). In *Methods in enzymology* (ed. D. V. Goeddel), Vol. 185, pp. 341–51. Academic Press, London.
9. Tuite, M. F., Dobson, M. J., Roberts, N. A., King, R. M., Burke, D. C., Kingsman, S. M., and Kingsman, A. J. (1982). *EMBO J.*, **1**, 603.
10. Johnston, S. A., Salmeron, J. M., and Dincher, S. S. (1987). *Cell*, **50**, 143.
11. Price, V. L., Taylor, W. E., Clevenge, W., Worthington, M., and Young, E. T. (1990). In *Methods in enzymology* (ed. D. V. Goeddel), Vol. 185, pp. 308–18. Academic Press, London.

12. Hinnen, A., Meyhack, B., and Heim, J. (1989). In *Yeast genetic engineering* (ed. P. J. Barr, A. J. Brake, and P. Valenzuela), pp. 193–213. Butterworths.

13. Etcheverry, T. (1990). In *Methods in enzymology* (ed. D. V. Goeddel), Vol. 185, pp. 319–29. Academic Press, London.

14. Brake, A. J., Merryweather, J. P., Coit, D. G., Heberlein, U. A., Masiarz, F. R., Mullenbach, G. T., Urdea, M. S., Valenzuela, P., and Barr, P. J. (1984). *Proc. Natl. Acad. Sci. USA*, **81**, 4642.

15. Walton, E. F. and Yarranton, G. T. (1989). In *Molecular and cell biology of yeasts* (ed. E. F. Walton and G. T. Yarranton), pp. 43–69. Blackie and Van Nostrand Reinhold.

16. Bitter, G. A. and Egan, K. M. (1988). *Gene*, **69**, 193.

17. Cousens, D. J., Wilson, M. J., and Hinchcliffe, E. (1990). *Nucleic Acids Res.*, **18**, 1308.

18. Cousens, L. S., Shuster, J. R., Gallegos, C., Ku, L., Stempien, M. M., Urdea, M. S., Sanchez-Pescador, R., Taylor, A., and Tekamp-Olson, P. (1987). *Gene*, **61**, 265.

19. Schena, M., Picard, D., and Yamamoto, K. R. (1991). In *Methods in enzymology* (ed. C. Guthrie and G. R. Fink), Vol. 194, pp. 389–98. Academic Press, London.

20. Purvis, I. J., Chotai, D., Dykes, C. W., Lubahn, D. B., French, F. S., Wilson, E. M., and Hobden, A. N. (1991). *Gene*, **106**, 35.

21. Pichuantes, S., Babé, L. M., Barr, P. J., and Craik, C. S. (1989). *Proteins: Struct. Funct. Genet.*, **6**, 324.

22. Ballou, C. E. (1970). *J. Biol. Chem.*, **245**, 1197.

23. Perlman, D., Halvorson, H. O., and Cannon, L. E. (1982). *Proc. Natl. Acad. Sci. USA*, **79**, 781.

24. Hitzeman, R. A., Chen, C. Y., Dowbenko, D. J., Renz, M. E., Lui, C., Pai, R., Simpson, N. J., Kohr, W. J., Singh, A., Chisholm, V., Hamilton, R., and Chang, C. N. (1990). In *Methods in enzymology* (ed. D. V. Goeddel), Vol. 185, pp. 421–40. Academic Press, London.

25. Chen, W. and Struhl, K. (1985). *EMBO J.*, **4**, 3273.

26. Rudolph, H. K., Antebi, A., Fink, G. R., Buckley, C. M., Dorman, T. E., Le Vitre, J., Davidow, L. S., Mao, J., and Moir, D. T. (1989). *Cell*, **58**, 133.

27. Tsai, P.-K., Frevert, J., and Ballou, C. E. (1984). *J. Biol. Chem.*, **259**, 3805.

28. Raschke, W. C., Kern, K. A., Antalis, C., and Ballou, C. E. (1973). *J. Biol. Chem.*, **248**, 4660.

29. Ford, C. F., Suominen, I., and Glatz, C. E. (1991). *Protein Expression and Purification*, **2**, 95.

30. Sambrook, J., Fritsch, E. F., and Maniatis, T. (ed.) (1989). *Molecular cloning, a laboratory manual*. Cold Spring Harbor Press, Cold Spring Harbor, NY.

31. Romanos, M. A., Makoff, A. J., Fairweather, N. F., Beesley, K. M., Slater, D. E., Rayment, F. B., Payne, M. M., and Clare, J. J. (1991). *Nucleic Acids Res.*, **19**, 1461.

32. Bennetzen, J. L. and Hall, B. D. (1982). *J. Biol. Chem.*, **257**, 3026.

33. Demoulder, J., Fiers, W., and Contreras, R. (1992). *Gene*, **111**, 207.

34. Kniskern, P. J., Hagopian, A., Montgomery, D. L., Burke, P., Dunn, N. R., Hofmann, K. J., Miller, W. J., and Ellis, R. W. (1986). *Gene*, **46**, 135.

35. Clare, J. J., Romanos, M. A., Rayment, F. B., Rowedder, J. E., Smith, M. A., Payne, M. M., Sreekrishna, K., and Henwood, C. A. (1991). *Gene*, **105**, 205.

36. Sleep, D., Belfield, G. P., and Goodey, A. R. (1990). *Bio/Technology*, **8**, 42.
37. Sleep, D., Belfield, G. P., Ballance, D. J., Steven, J., Jones, S., Evans, L. R., Moir, P. D., and Goodey, A. R. (1991). *Bio/Technology*, **9**, 183.
38. Shuster, J. R., Moyer, D. L., Lee, H., Dennis, A., Smith, B., and Merryweather, J. P. (1989). *Gene*, **83**, 47.
39. Guthrie, C. and Fink, G. R. (ed.) (1991). *Methods in enzymology*, Vol. 194. Academic Press, London.
40. Barr, P. J., Brake, A. J., and Valenzuela, P. (ed.) (1989). *Yeast genetic engineering*. Butterworths.

6

Interaction trap cloning with yeast

RUSSELL L. FINLEY JR and ROGER BRENT

1. Introduction

The interaction trap is a two-hybrid system for cloning cDNAs encoding proteins that interact with a protein whose coding sequences are known. The method uses the transcription of yeast reporter genes as a synthetic phenotype to detect protein–protein interactions. It can also be used to study interactions between known proteins.

1.1 Background

The two-hybrid approach takes advantage of the modular domain structure of eukaryotic transcription factors. Many eukaryotic transcription activators have at least two distinct functional domains, one that directs binding to specific DNA sequences and one that activates transcription (1, 2). This modular structure is best illustrated by yeast experiments showing that the DNA binding domains or activation domains can be exchanged from one transcription factor to the next and retain function. For example, when the DNA binding domain of the yeast transcription factor Gal4 is replaced with the DNA binding domain of the bacterial repressor LexA, the resulting hybrid protein activates transcription of genes containing upstream LexA binding sites (3). Similarly, when the DNA binding domain of Gal4, which by itself does not activate transcription, is fused to activation domains from other proteins the resulting hybrid proteins activate transcription of reporters with upstream Gal4 binding sites (4–6). A crucial corollary of the modular nature of transcription activators is that the DNA binding and activation domains need not be covalently attached to each other for activation to occur. This was first demonstrated by Ma and Ptashne (7) with a Gal4 derivative that contained the DNA binding domain as well as a domain that interacts with another yeast protein, Gal80, but that lacked the activation domain. When this derivative was expressed in yeast it did not activate transcription of a reporter gene containing upstream Gal4 binding sites. However, when it was co-expressed with a second, hybrid protein, consisting of Gal80 fused to an activation domain, interaction between the Gal4 DNA binding derivative and the Gal80 activation domain hybrid resulted in activation of the reporter gene.

The general utility of the modularity of transcription factors was demonstrated by Fields and Song (8) who showed yeast transcription could be used to assay the interaction between two proteins if one of them was fused to a DNA binding domain and the other was fused to an activation domain. In their experiment, one of the hybrid proteins contained the DNA binding domain of Gal4 fused to the yeast protein Snf1, and the other contained the activation domain of Gal4 fused to another yeast protein, Snf4. When Snf1 and Snf4 interacted they brought together the DNA binding and activation domains, so that the two hybrid proteins bound to Gal4 binding sites upstream of a *lacZ* reporter gene and activated its transcription. Thus, the interaction between Snf1 and Snf4 was assayed as production of β-galactosidase. The success of this experiment prompted Fields and Song to suggest that yeast transcription could be used in this way to clone cDNAs encoding proteins that interact with a given known protein (8). In their scheme, a known protein is expressed fused to the DNA binding domain of Gal4, and a cDNA library is expressed so that proteins encoded by the cDNA are fused to an activation domain (activation-tagged). Transcription of a reporter gene will be activated in yeast containing activation-tagged cDNA encoded proteins that interact with the known protein.

Based on this suggestion, two-hybrid cloning systems have been developed in several laboratories (9–13). All have three basic components:

- yeast vectors for expression of a known protein fused to a DNA binding domain
- yeast vectors that direct expression of cDNA encoded proteins fused to a transcription activation domain
- yeast reporter genes that contain binding sites for the DNA binding domain.

These components differ in detail from one system to the other. All systems utilize the DNA binding domain from either Gal4 or LexA. The Gal4 domain is efficiently localized to the yeast nucleus where it binds with high affinity to well-defined binding sites which can be placed upstream of reporter genes (14–16). LexA does not have a nuclear localization signal, but enters the yeast nucleus and, when expressed at a sufficient level, efficiently occupies LexA binding sites (operators) placed upstream of a reporter gene (3, 17, 18). No endogenous yeast proteins bind to the LexA operators. Different systems also utilize different reporters. Most systems use a reporter that has a yeast promoter, either from the *GAL1* gene or the *CYC1* gene, fused to *lacZ* (19, 20). These *lacZ* fusions either reside on multicopy yeast plasmids or are integrated into a yeast chromosome. To make the *lacZ* fusions into reporters, the *GAL1* or *CYC1* transcription regulatory regions have been removed and replaced with binding sites that are recognized by the DNA binding domain being used. A screen for activation of the *lacZ* reporters is performed by plating yeast on indicator plates that contain X-Gal (5-bromo-4-chloro-3-indolyl-β-D-galactopyranoside); on this medium, those yeast colonies in which the reporters are transcribed produce β-galactosidase and turn blue.

Some systems use a second reporter gene and a yeast strain that requires expression of this reporter to grow on a particular medium. These 'selectable marker' genes usually encode enzymes required for the biosynthesis of an amino acid. Such reporters have the marked advantage of providing a selection for cDNAs that encode interacting proteins, rather than a visual screen for blue yeast. To make appropriate reporters from the marker genes, their upstream transcription regulatory elements have been replaced by binding sites for a DNA binding domain. The *HIS3* and *LEU2* genes have both been used as reporters in conjunction with appropriate yeast strains that require their expression to grow on media lacking either histidine or leucine, respectively.

Finally, different systems use different means to express activation-tagged cDNA proteins. In all current schemes, the cDNA encoded proteins are expressed with an activation domain at the amino terminus. The activation domains used include the strong activation domain of Gal4, the very strong activation domain of the viral protein VP16, or a weaker activation domain derived from bacteria, called B42. The activation-tagged cDNA encoded proteins are expressed either from a constitutive promoter, or from a conditional promoter such as that of the *GAL1* gene. Use of a conditional promoter makes it possible to quickly demonstrate that activation of the reporter gene is dependent on expression of the activation-tagged cDNA proteins.

Many of these systems now provide the investigator with a relatively good chance to recover proteins that interact with other proteins. Because most are based on the same concepts, some of their components are often interchangeable. However, different systems utilize the yeast selectable markers in different ways. Moreover, systems that employ the DNA binding domain of Gal4 must use a yeast strain that lacks wild-type Gal4; these systems cannot use library vectors that direct synthesis of the activation-tagged cDNA encoded proteins from the *GAL1* promoter whose transcription requires Gal4.

1.2 The interaction trap

The interaction trap is an implementation of the two-hybrid system developed by Gyuris *et al.* (11). It consists of three critical components. First, it uses a vector for expression of a protein of interest fused to LexA. Because the goal of interaction trap cloning is to find proteins that interact with the protein fused to LexA, this hybrid is referred to as the 'bait'. Secondly, it uses a yeast strain with two reporter genes. One reporter is a yeast *LEU2* derivative that has its normal upstream regulatory sequences replaced with LexA operators.

Transcription of the LexA–operator–*LEU2* gene (*LexAop-LEU2*) can be measured by the ability of the strain to grow in the absence of leucine, which requires the *LEU2* gene product. The *LexAop-LEU2* gene is integrated into the yeast chromosome. The other reporter gene is *lacZ* which provides a secondary assay of activation by the bait and activation-tagged proteins interacting with it, as well as some quantitative information about the interaction. Thirdly, it uses a library plasmid that directs the conditional expression of cDNA encoded proteins fused at their amino termini to a moiety containing three domains: a nuclear localization signal, a transcription activation domain, and an epitope tag. The activation-tagged cDNA encoded protein is expressed from the yeast *GAL1* promoter, which is induced by galactose and repressed by glucose.

The interaction trap is illustrated in *Figure 1*. The bait protein is constitutively expressed from a plasmid with the *HIS3* marker. It binds to LexA operators upstream of the reporter genes, *LEU2* and *lacZ*, but does not activate transcription. The activation-tagged cDNA encoded protein is conditionally expressed from the *GAL1* promoter. In glucose medium (*Figure 1a*) the *GAL1* promoter is repressed, no cDNA encoded protein is made, and the yeast does not grow in the absence of leucine. When the yeast are grown on

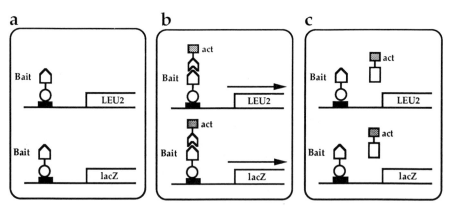

Figure 1. The interaction trap. (a) Glucose medium. The LexA fusion protein (bait) is made and binds to LexA operators (black box) upstream of the two reporter genes, *LEU2* and *lacZ*. The bait does not activate transcription of the reporters. The activation-tagged cDNA encoded protein is not expressed because the *GAL1* promoter on the library plasmid is repressed in the presence of glucose. The cell does not form a colony on a medium lacking leucine and forms a white colony on an X-Gal plate. (b) Galactose medium: interaction. Here, galactose induces expression of an activation-tagged cDNA encoded protein that interacts with the bait. The activation domain activates transcription of *LEU2* and *lacZ*. The cell forms a colony on a medium lacking leucine and forms a colony that turns blue on an X-Gal plate. (c) Galactose medium: no interaction. Here, galactose induces expression of an activation-tagged cDNA encoded protein that does not interact with the bait. The cell does not grow on a medium lacking leucine and forms a white colony on X-Gal plates.

Table 1. Flow chart of an interactor hunt

1. Construct the bait plasmid and introduce it into a yeast strain containing the *lacZ* reporter plasmid and the integrated *LexAop-LEU2* gene. This creates the selection strain.
2. Show that the bait does not activate the reporter genes on its own.
3. Using a different, repression strain, verify that the bait is expressed, enters the nucleus, and binds to reporter genes.
4. Introduce the cDNA library plasmid into the selection strain.
5. Plate transformants on to galactose plates lacking leucine.
6. Pick Leu$^+$ colonies and transfer to glucose master plates (containing leucine) to shut off synthesis of the activation-tagged cDNA encoded protein.
7. Determine which of the picked colonies exhibit galactose-dependent activation of both reporter genes by replica plating from the glucose master plates to two different plates that contain either glucose or galactose but lack leucine, and two different plates that contain either glucose or galactose and also contain X-Gal.
8. Show that the colonies that are galactose-dependent Leu$^+$ and lacZ$^+$ contain activation-tagged cDNA proteins that interact specifically with the bait. Do this by isolating the library plasmids and re-introducing them into the original bait strain and into other strains expressing unrelated baits.

galactose medium (*Figure 1b* and *c*), activation-tagged cDNA encoded proteins are expressed. Those that interact with the bait activate transcription of the *LEU2* and *lacZ* reporters (*Figure 1b*). Thus, cells containing activation-tagged cDNA proteins that interact with the bait form colonies on galactose medium lacking leucine and form blue colonies on galactose X-Gal plates. Cells containing activation-tagged cDNA encoded proteins that do not interact with the bait cannot grow on galactose medium lacking leucine and form white colonies on X-Gal plates (*Figure 1c*).

An outline of an interactor hunt is presented in *Table 1*. The protocols for using the interaction trap described below require knowledge of a few basic yeast microbiological and genetic techniques. A more detailed description of such techniques, together with recipes for appropriate media, can be found elsewhere (21–23).

2. Making and testing baits

2.1 LexA fusion expression plasmids

To make a plasmid that directs the synthesis of the LexA fusion or 'bait' protein, the coding region for the protein of interest is inserted into pEG202 (*Figure 2*) or a related plasmid (11). pEG202 is a multicopy yeast plasmid containing the yeast 2μ origin of replication and the selectable marker gene *HIS3*, as well as the full-length LexA coding region flanked by the *ADH1* promoter and terminator. Bait proteins expressed from this plasmid contain

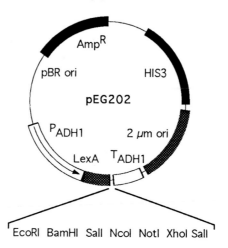

Figure 2. pEG202. pEG202 (11) is a yeast–*E. coli* shuttle vector that contains a yeast expression cassette that includes the promoter from the yeast *ADH1* gene (P_{ADH1}), sequences that encode amino acids 1–202 of the bacterial repressor protein LexA, a polylinker, and the transcription terminator sequences from the yeast *ADH1* gene (T_{ADH1}). It also contains a *E. coli* origin of replication (pBR ori), the ampicillin resistance gene (Amp^R), a yeast selectable marker gene (HIS3), and a yeast origin of replication (2 μm ori). pEG202 confers upon a *his3*$^-$ yeast strain the ability to grow in the absence of histidine and directs the constitutive expression of LexA (fused to approximately 17 amino acids encoded by the polylinker). Protein coding sequences can be inserted in-frame with LexA into the unique restriction sites shown.

amino acids 1–202 of LexA, which include the DNA binding and dimeriza-tion domains. Downstream of the LexA coding region in pEG202 are unique *Eco*RI, *Bam*HI, *Sal*I, *Nco*I, *Not*I, and *Xho*I cloning sites. The bait plasmid can be introduced and maintained in a *his3* yeast strain (e.g. EGY48, see below) by selecting transformants on media lacking histidine. Transformants will constitutively express the protein of interest with LexA at its amino terminus. Although it does not contain a yeast nuclear localization signal, LexA and most LexA fusions will enter the nucleus (3, 17, 18, 24–29). The expression levels afforded by the *ADH1* promoter are generally sufficient to provide occupancy of LexA operators upstream of the reporter genes. For the rare bait that is excluded from the nucleus, a pEG202 derivative can be used that directs expression of LexA fusions that contain a nuclear localization signal (W. Breitwieser and A. Ephrussi, personal communication).

2.2 Reporters and yeast strains

2.2.1 *LEU2* reporter strains

An interactor hunt employs a selection for cDNAs encoding interactors. This is made possible by use of a yeast strain, EGY48 (11), that has an integrated

LEU2 gene with its upstream regulatory region replaced by LexA operators. The strain has no other *LEU2* gene and does not grow in the absence of leucine unless the *LexAop-LEU2* gene is transcribed. The *LEU2* reporter in EGY48 is very sensitive; it is activated by even weak transcription activators fused to LexA, or by activation-tagged proteins that interact weakly with LexA fusions. The high sensitivity is due to the presence of three high affinity LexA operators positioned near the *LEU2* transcription start. The operators are from the bacterial *colE1* gene and each can potentially bind two LexA dimers (30).

While the sensitivity of EGY48 is an advantage in that it facilitates isolation of activation-tagged cDNA proteins that interact weakly with the bait, this strain may be too sensitive for use with baits that are themselves weak transcription activators. Many proteins, including some that are not transcription factors, will activate transcription of *LEU2* in EGY48. For a bait to be used in an interactor hunt it must not activate *LEU2* transcription. For baits that fail to meet this criterion, two approaches can be taken:

(a) The sensitivity of the reporter strain can be reduced. One way to do this is by using a strain containing fewer operators upstream of *LEU2* (e.g. one operator instead of three) (E. Golemis, D. Krainc, R. L. Finley, unpublished data). If a bait still activates transcription of *LEU2* in a strain with only one operator, the sensitivity can be reduced further by using a diploid yeast strain, in which, for unknown reasons, LexA fusions activate less transcription of the reporter genes (E. Golemis, A. Mendelsohn, D. Krainc, R. L. Finley, unpublished data).

(b) A more drastic approach when using a bait that activates transcription is to construct deletion derivatives of it that do not activate. A good way to start, if prior knowledge of the precise location of transcription activation domains is unavailable, is to construct derivatives that lack highly acidic regions which are often responsible for transcription activation in yeast (2, 4, 31). The obvious disadvantage of this approach is that regions important for interactions with other proteins may be removed.

In addition to the mutation in the endogenous *LEU2* gene, EGY48 and related strains carry mutations in three other marker genes (*his3*, *trp1*, *ura3*) that are needed to allow selection of the plasmids used in the interaction trap. The *his3* mutation is complemented by the *HIS3* gene on the bait expression vector. The *trp1* and *ura3* mutations are respectively complemented by the *TRP1* gene on the library plasmid and the *URA3* gene on the *lacZ* reporter plasmid (see below). The plasmids for the bait, library, and *lacZ* reporter each contain the yeast 2μ origin of replication so that under continued selection they should be maintained at 20–100 copies per cell (32).

2.2.2 LacZ reporters

In addition to the *LexAop-LEU2* reporter, an interactor hunt employs a *LexAop-lacZ* reporter. The *lacZ* reporters reside on 2μ plasmids containing the *URA3* gene and the *GAL1* TATA, transcription start, and a small part of the *GAL1* coding sequence fused to *lacZ* (19, 33). The *GAL1* upstream activating sequences (UAS$_G$) have been replaced with an *Xho*I site into which various numbers of LexA operators have been inserted (see *Figure 3*). In the absence of a LexA fusion, or interacting activation-tagged protein, yeast bearing these reporters make no detectable β-galactosidase and appear white on X-Gal plates. Use of the *lacZ* reporters provides two advantages in an interactor hunt:

(a) False positives that may arise by activation of the *LEU2* gene due to a yeast mutation, or to binding of the activation-tagged cDNA protein to the *LEU2* promoter, can be identified because they will fail to activate the *lacZ* reporter.

(b) The *lacZ* reporters provide a relative measure of the amount of transcription caused by the interaction of an activation-tagged cDNA protein with a bait.

The phenotype measured with the *LEU2* reporter, growth in the absence of leucine, while very sensitive, is difficult to quantitate. In contrast, the β-galactosidase activity in a yeast is directly proportional to the amount of *lacZ* transcription, and is easily measured (33). Careful use of the *lacZ* reporters may even allow comparison of interaction affinities between different baits and activation-tagged proteins (11).

The threshold affinity of protein–protein interactions to be detected in an interactor hunt can be adjusted by choosing between different *lacZ* reporters. The sensitivity of the *lacZ* reporter phenotype depends on the number of LexA operators positioned upstream of *lacZ*. Activation-tagged proteins that interact weakly with a bait can be identified by using a more sensitive *lacZ* reporter containing more operators. However, if too many false positives are obtained with a particular bait, the search can be limited in order to find only cDNA encoded proteins that interact tightly with the bait by using a less sensitive *lacZ* reporter. All *lacZ* reporters commonly used are less sensitive than *LexAop-LEU2* reporters. Because of this, some LexA activators will activate *LEU2* and allow EGY48 to grow in the absence of leucine, but will not activate *lacZ* and cause the strain to turn blue on X-Gal plates.

2.3 Testing the bait protein

Before conducting an interactor hunt, the bait should be tested to ensure that the fusion protein enters the nucleus, binds LexA operators, and does not activate transcription of the *LEU2* gene. This is done in two steps:

(a) A selection strain is made by introducing the bait plasmid into EGY48 that contains a *LexAop-lacZ* reporter. The selection strain is used to show that the bait protein does not activate transcription of *LEU2* and *lacZ* (*Protocol 1*). Eventually, the library will be introduced into this strain.

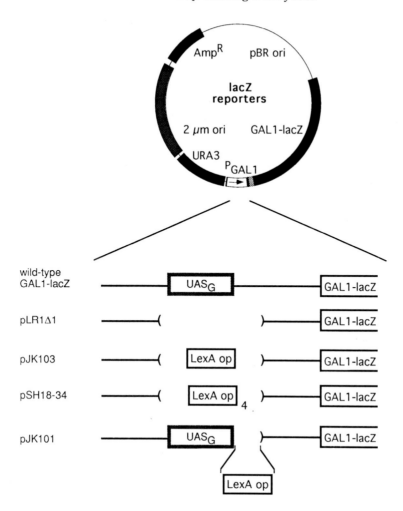

Figure 3. *LacZ* reporters. The *lacZ* reporters are derived from a plasmid that contains the wild-type *GAL1* promoter fused to *lacZ* (19). Reporters for measuring activation are derived from pLR1△1, in which the *GAL1* upstream activation sequences (UAS$_G$) have been deleted (33). Various numbers and types of LexA operators have been inserted in place of UAS$_G$ to create *lacZ* reporters with different sensitivities. Two examples are shown: pJK103 contains a single high affinity overlapping type *colE1* LexA operator which binds two LexA dimers; pSH18–34 contains four of these *colE1* operators and so is much more sensitive than pJK103 to activation by LexA fusions and activation-tagged proteins that interact with them. Derivatives with two, three, or more than four *colE1* operators have also been made (S. Hanes, personal communication) (28). A reporter less sensitive than pJK103, pRB1840, contains one lower affinity LexA operator (3). pJK101 is used to measure repression by LexA fusions (18, 29). It contains most of UAS$_G$ and one *colE1* operator between UAS$_G$ and the *GAL1* TATA. All of the *lacZ* reporters also contain a *E. coli* origin of replication (pBR ori), the ampicillin resistance gene (Amp^R), a yeast selectable marker gene (URA3), and a yeast origin of replication (2 μm ori).

177

(b) The bait plasmid is introduced into a related strain that contains a different *lacZ* reporter to verify that the bait protein enters the yeast nucleus and binds LexA operators. This is done with a repression assay (*Protocol 2*).

2.3.1 Testing whether the bait protein activates transcription of the reporters

Protocol 1 describes how to verify that the bait does not activate transcription of the *LEU2* reporter. In addition to the bait expression plasmid, this protocol uses three related *HIS3* plasmids as controls. The first is a plasmid that makes no LexA protein or that makes LexA fused to a protein that does not activate transcription. EGY48 derivatives that contain such plasmids fail to grow on media lacking leucine. The second is a plasmid that makes LexA fused to a transcription activation domain, like the activation domain of Gal4. Such a plasmid will allow EGY48 to grow in the absence of leucine. The third is the parent plasmid, pEG202. The LexA protein encoded by pEG202 includes several amino acids encoded by the polylinker which make the protein a weak transcription activator. EGY48 containing pEG202 grow slowly on media lacking leucine. A good criterion for determining whether a bait plasmid can be used for an interactor hunt is to show that it causes EGY48 to grow more slowly on media lacking leucine than pEG202.

Only a handful of yeast transformants are needed in *Protocol 1* and in the repression assay (*Protocol 2*; see later). However, the transformation efficiency (transformants per microgram of plasmid) must be very high for *Protocol 3* (Section 4.1.1), in which the selection strain is transformed with the library. To become familiar with high efficiency yeast transformation, it is therefore advisable to use a high efficiency method for the transformations in *Protocols 1* and *2*. There are several effective high efficiency yeast transformation protocols to choose from including electroporation (34, 35) and those that employ lithium salts (36). The method described in *Protocol 3* (Section 4.1.1) results in about 10^5 transformants per microgram and may be scaled down for use in *Protocols 1* and *2*.

Two other considerations regarding yeast transformation deserve mention. First, once a plasmid has been introduced into a strain it must be maintained by continued selection for its presence. Thus, a strain that already contains the *URA3 lacZ* reporter plasmid should be transformed with a second plasmid by first growing on media lacking uracil (Glu ura⁻; see *Table 2* for media designations). Although strains that contain different plasmids can be constructed by introducing more than one plasmid at a time, the transformation efficiency will be lower than when strains are constructed by serial transformation of one plasmid at a time. Second, each time a strain is transformed, a control transformation should be performed using no plasmid DNA.

Table 2. Media recipes

Dropout media
Dropout media, also known as complete minimal (CM) dropout media, contains a nitrogen base, a mixture of nutrients shown below with one or a few left out (dropped out), and a carbon source (usually a sugar). It is convenient to make a dropout powder corresponding to each of the dropout media that will be used. Thus, for media lacking tryptophan (trp⁻), the dropout powder used would contain all of the nutrients listed below except for tryptophan. To make the dropout powder, combine all but the appropriate nutrients, e.g. if making media lacking tryptophan (trp⁻) leave out tryptophan. Grind all of the components of the dropout powder in a clean dry mortar and pestle until homogeneous and store at room temperature. Three separate stocks of the carbon sources (20% galactose, 20% glucose, 20% raffinose) should also be made and filter sterilized.

Nutrient	Amount in dropout powder (g)	Final concentration in medium ($\mu g/ml$)
Adenine	2.5	40
L-Arginine (HCl)	1.2	20
L-Aspartic acid	6.0	100
L-Glutamic acid (monosodium)	6.0	100
L-Histidine (his)	1.2	20
L-Isoleucine	1.8	30
L-Leucine (leu)	3.6	60
L-Lysine	1.8	30
L-Methionine	1.2	20
L-Phenylalanine	3.0	50
L-Serine	22.5	375
L-Threonine	12.0	200
L-Tryptophan (trp)	2.4	40
L-Tyrosine	1.8	30
L-Valine	9.0	150
Uracil (ura)	1.2	20

Dropout plates
For 1 litre, mix in 850 ml deionized H_2O:
- 6.7 g yeast nitrogen base (YNB) without amino acids (Difco)
- 2 g dropout powder lacking the appropriate nutrient(s) (see above)
- one pellet of NaOH (~ 0.1 g)
- 20 g agar (Difco Bacto agar)

Autoclave. Add the appropriate carbon source from sterile 20% stocks. For Gal/Raf plates add galactose to 2% and raffinose to 1% final concentrations; for Glu plates add glucose to 2% final concentration.

Liquid dropout media
Make liquid dropout media the same way as dropout plates (above), leaving out the agar and NaOH pellet.

Table 2. *continued*

YPD media (also known as YEPD)
YPD plates
For 1 litre, mix in 900 ml deionized H_2O:
- 10 g yeast extract (Difco Bacto yeast extract)
- 20 g peptone (Difco Bacto peptone)
- one pellet of NaOH ($\sim$ 0.1 g)
- 20 g agar (Difco Bacto agar)
Autoclave. Add 100 ml sterile 20% glucose.

Liquid YPD medium
Make liquid YPD medium the same way as YPD plates (above), leaving out the agar and NaOH pellet.

X-Gal plates
For 1 litre, mix in 800 ml deionized H_2O:
- 6.7 g YNB without amino acids
- 1.5 g dropout powder
- 20 g agar (Difco Bacto agar)
Autoclave. Immediately add the appropriate sterile 20% carbon source(s).
Allow to cool to 65°C.[a]
- add 100 ml of 10 × BU salts (see below)
- add 4 ml of 20 mg/ml X-Gal dissolved in dimethylformamide (stored at −20°C)

> *10 × BU salts*
> For 1 litre
> - 70 g $Na_2HPO_4 \cdot 7H_2O$
> - 30 g NaH_2PO_4
> Adjust pH to 7.0. Autoclave, store at room temperature.

[a] Adding salts while medium is too hot causes the salts to precipitate. Also, X-Gal is thermolabile and will be destroyed if added to hot media.

Protocol 1. Testing baits for transcription activation

Reagents

- Liquid YPD media (see *Table 2*)
- Liquid dropout media (Glu ura⁻, Glu ura⁻his⁻; see *Table 2*)
- Dropout plates (Glu ura⁻, Glu ura⁻his⁻, Gal/Raf ura⁻his⁻, Gal/Raf ura⁻his⁻leu⁻; see *Table 2*)
- X-Gal plates (Glu ura⁻his⁻ X-Gal; see *Table 2*)

Method

1. To construct the selection strain, transform[a] EGY48 (grown in liquid YPD medium) with a *URA3 lacZ* reporter plasmid and select transformants on Glu ura⁻ plates.

2. Combine three colonies from the Glu ura⁻ plates, grow in Glu ura⁻ liquid, and transform them with the *HIS3* bait plasmid (or control plasmids). Select transformants on Glu ura⁻his⁻ plates.

3. Pick four individual colonies from each transformation and use each colony to inoculate 5 ml Glu ura⁻his⁻ liquid cultures. At the same time, streak the same four transformants to another Glu ura⁻his⁻ plate for storage and later recovery. All four should behave identically in the tests below, in which case any one will serve as the selection strain into which the library will be introduced. Incubate these plates at 30 °C until colonies form (about two days) and then store at 4 °C.

4. Grow the liquid cultures at 30 °C shaking to OD_{600} = 0.5 (corresponding to about 10^7 cells/ml). This is mid-log phase. If the overnight cultures grow to a density greater than OD_{600} = 0.5, dilute to less than OD_{600} = 0.2 and then grow to OD_{600} = 0.5 so that the cells are in mid-log phase when harvested.

5. Make 10^2- and 10^3-fold dilutions of each culture in sterile water.

6. Spot 10 µl of the culture and 10 µl of the dilutions onto two plates[b]:
   • Gal/Raf ura⁻his⁻
   • Gal/Raf ura⁻his⁻leu⁻.

   Spot yeast containing the bait plasmid being tested and yeast containing the control plasmids on the same plates for side-by-side comparison. Incubate at 30 °C.

7. Monitor the growth of the cells in the spots for several days. The yeast in all of the spots should grow at a similar rate on the Gal/Raf ura⁻his⁻ plates. If growth on the Gal/Raf ura⁻his⁻ plates is reproducibly diminished for a given bait, relative to the controls, it may indicate that its expression is toxic to yeast. Yeast with no LexA or with a non-activating LexA fusion should not grow after several days on Gal/Raf ura⁻his⁻leu⁻ plates. Yeast containing LexA fused to a protein that activates transcription may grow as fast on Gal/Raf ura⁻his⁻leu⁻ plates as on Gal/Raf ura⁻his⁻ plates, depending on the strength of the activator.

   The above steps establish whether a bait can be used in an interactor hunt. For a bait to be used, the selection strain containing it must not grow on Gal/Raf ura⁻his⁻leu⁻ plates for two to three days. If yeast that contain a bait begin to form visible colonies before three days, it may be necessary to construct a less sensitive selection strain (see Section 2.2.1). However, if the yeast grow very slowly (e.g. form colonies only after three days) it is still possible to do an interactor hunt by selecting yeast with library plasmids that cause colonies to form in one or two days.

8. Look for *lacZ* expression in the selection strain. Patch individual

Protocol 1. *Continued*

transformants from step 2 to Glu ura⁻ his⁻ X-Gal plates and incubate at
30 °C. Yeast with the control LexA–activator fusion should turn blue
overnight while those lacking LexA or containing a transcriptionally
inert bait will remain white after many days. If a bait activates the *LEU2*
gene but not the *lacZ* gene (which is frequently observed because the
LexAop-LEU2 reporter is more sensitive than *LexAop-lacZ* reporter), it
may be possible to perform an interactor hunt by screening for yeast
containing activation-tagged cDNA proteins that interact with the bait
and activate the *lacZ* reporter; these yeast will be blue on X-Gal plates.
However, this method loses the advantage of a selection.

ᵃ See text in Section 2.3.1 for comments on transformation.
ᵇ Galactose is used in the media because the actual selection will eventually be done on
galactose plates to induce expression of the activation-tagged cDNA protein. Raffinose is added
to aid yeast growth; it provides a better carbon source than galactose alone but does not block
the ability of galactose to induce the *GAL1* promoter (R. L. Finlay, unpublished data).

2.3.2 Demonstrating that the bait enters the yeast nucleus and binds operators

Baits that do not activate the *LEU2* reporter in the assay in *Protocol 1* should
be tested to be sure they enter the nucleus and bind to LexA operators. This
can be done with a repression or 'blocking' assay. The repression assay is
based on the observation that LexA and non-activating LexA fusions can
repress transcription of a yeast reporter gene that has operators positioned
between the TATA and upstream activating sequence (UAS) (18). The
mechanism of this repression is not understood but presumably is not equivalent to repression by repressor proteins native to yeast (37). While some
LexA fusions repress more than others (24), repression does depend on the
presence of operators in the reporter, and thus any repression observed may
be attributed to operator occupancy by the bait.

The reporter plasmid used for the repression assay (pJK101; see *Figure 3*) is
similar to the plasmids used to test activation; it contains the 2μ origin,
URA3, and a *GAL1–lacZ* fusion. Unlike the plasmids used for testing activation, the *GAL1–lacZ* fusion in pJK101 contains most of the *GAL1* upstream
activating sequence, UAS_G. In addition it contains one LexA operator positioned between UAS_G and the TATA box. *LacZ* expression is induced by
galactose and is detectable in the presence of glucose, because negative
regulatory elements that normally keep *GAL1* completely repressed in
glucose are not present (38). Transcriptionally inert LexA fusions that bind to
the operator in pJK101 repress *lacZ* expression from 2- to 20-fold in the
presence of galactose. Repression appears more profound when the yeast are
grown in glucose medium because there is less *lacZ* expression to begin with.
The repression assay can often be done on X-Gal plates by looking for

differences in blueness between yeast with different baits. However, when looking for low levels of *lacZ* expression in yeast grown in glucose, or when looking for slight differences in *lacZ* expression (two- to fourfold), more sensitive β-galactosidase assays may be necessary (39, 40).

Protocol 2. The repression assay

Reagents

- Liquid YPD media (see *Table 2*)
- Liquid dropout media (Glu ura⁻, Glu ura⁻ his⁻; see *Table 2*)
- Dropout plates (Glu ura⁻, Glu ura⁻ his⁻; see *Table 2*)
- X-Gal plates (Glu ura⁻his⁻ X-Gal, Gal/Raf ura⁻his⁻ X-Gal; see *Table 2*)

Method

1. Transform[a] EGY48 with pJK101 and select transformants on Glu ura⁻ plates.

2. Combine three colonies from these plates and transform[a] them with the *HIS3* bait plasmid (or *HIS3* control plasmids). Select transformants on Glu ura⁻his⁻ plates.

3. Pick four individual colonies from each transformation and streak a patch of them on to Glu ura⁻his⁻ and Gal/Raf ura⁻his⁻ plates containing X-Gal. Incubate at 30 °C.

4. Examine the X-Gal plates after one, two, and three days. Yeast lacking LexA will begin to turn blue on the Gal/Raf X-Gal plates after one day and will appear light blue on the Glu X-Gal plates after two or more days. Yeast containing a bait that enters the nucleus and binds operators turn blue more slowly than the yeast lacking LexA.

5. Baits that repress transcription of *lacZ* in pJK101 by twofold or less may not cause a visible reduction in blue on X-Gal plates. If no repression is observed on the X-Gal plates, perform β-galactosidase assays with transformants from step 2. Grow the transformants in 5 ml Glu ura⁻ his⁻ and Gal/Raf ura⁻his⁻ liquid media, or on Glu ura⁻his⁻ and Gal/Raf ura⁻his⁻ plates for two days, before doing β-galactosidase assays (39, 40).

[a] See text in Section 2.3.1 for comments on yeast transformation.

2.3.3 Verifying that a full-length fusion protein is made

Finally, it is usually good practice to demonstrate that the full-length bait protein is made. This can be done by running extracts from yeast cells that harbour the bait plasmid on an SDS–polyacrylamide gel, immunoblotting with either an antibody to LexA or one specific to the protein fused to LexA (27, 29), and detecting a fusion protein of the expected apparent molecular weight. Yeast cell extracts can be prepared by growing yeast in liquid culture (lacking histidine to maintain selection for the bait plasmid) to OD_{600} of 0.5, spinning 1 ml of the culture to pellet the cells, and resuspending the cells in 50 μl of 2 × Laemmli sample buffer (41). The cells can then be broken by freezing on dry ice followed by boiling for five minutes prior to loading on an SDS–polyacrylamide gel (about 15 μl/lane). The proteins can then be transferred to a filter and blotted with standard immunoblotting (Western) methods (22, 42).

3. Libraries

In an interactor hunt the expressed cDNA encoded proteins are fused to an activation domain, as well as a nuclear localization signal to increase their nuclear concentration, and an epitope tag so they may be immunologically identified. The prototypical library plasmid for expression of these activation-tagged cDNA proteins is pJG4–5 (11) (see *Figure 4*). This is a 2μ plasmid that contains the *TRP1* marker and the *GAL1* promoter. Downstream of the GAL1 promoter there is an ATG followed by 105 codons. These encode nine amino acids from the SV40 large T nuclear localization signal, 87 amino acids that make up the activation domain called B42, followed by nine amino acids comprising the haemagglutinin (HA) epitope tag. The B42 domain is derived from *E. coli* and acts as a moderately strong transcription activation domain in yeast (4). Use of this activation domain avoids the possible toxic effects of over-expressing a strong activation domain (43, 44). Downstream of the sequences that encode the fusion moiety there are unique *Eco*RI and *Xho*I sites for insertion of cDNAs. Proteins encoded by cDNAs that are in-frame carry this fusion moiety at their amino terminus. The activation-tagged cDNA encoded proteins will be expressed in yeast grown on galactose but not in yeast grown in glucose.

Numerous libraries have been made using pJG4–5. These include cDNA libraries made from RNA derived from HeLa cells (11), *Drosophila* ovaries, discs, and 0–12 hour embryos (R. L. Finley and R. Brent, in preparation), adult *Drosophila* heads (J. Huang and M. Rosbach, personal communication), *Drosophila* 16–26 hour embryos (V. Neel and M. Young, personal communication), serum-starved WI38 cells (C. Sardet, J. Gyuris, R. Brent, unpublished data), human brain (D. Krainc and R. Brent unpublished data), Arabidopsis (H. Zhang and H. Goodman, personal communication), and a library made from yeast genomic DNA (P. Watt, personal communication). Construction of libraries is a complex topic beyond the scope of this article, but is described elsewhere (11, 45, 46) (R. L. Finley and R. Brent, in preparation).

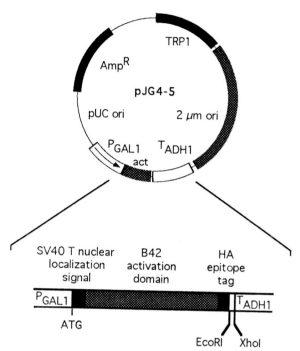

Figure 4. pJG4–5. pJG4-5 (11) is a yeast–*E. coli* shuttle vector that contains a yeast expression cassette that includes the promoter from the yeast *GAL1* gene (P_{GAL1}), followed by sequences that encode the 106 amino acid fusion moiety or activation tag, and the transcription terminator sequences from the yeast *ADH1* gene (T_{ADH1}). cDNAs or other protein coding sequences can be inserted into the unique *Eco*RI and *Xho*I sites so that encoded proteins are expressed with the fusion moiety at their amino terminus. The fusion moiety includes the nuclear localization signal from the SV40 virus large T antigen (PPKKKRKVA) (53), the B42 transcription activation domain (4), and the haemagglutinin (HA) epitope tag (YPYDVPDYA) (54). The plasmid also contains an *E. coli* origin of replication (pUC ori), the ampicillin resistance gene (Amp^R), a yeast selectable marker gene (TRP1), and a yeast origin of replication (2 μm ori).

4. An interactor hunt

4.1 Introducing the library into the selection strain

4.1.1 Two-step selection of interactors from library transformants

To conduct an interactor hunt, the library is introduced into the selection strain in a large transformation so that many transformed cells are obtained, each of which contains an individual library plasmid. Those cells that contain library plasmids encoding proteins that interact with the bait protein are then selected. To date, the most frequently successful method has been to do this in two steps:

(a) In the first step, yeast that already contain the bait and *lacZ* plasmids are transformed with the library and library transformants are isolated and frozen.

(b) In the second step, the selection for interactors is applied to the library transformants.

In this approach, yeast transformed with the *TRP1* library plasmid are selected by plating the transformation mix on to medium lacking tryptophan (it also lacks uracil and histidine to maintain selection for the bait and *lacZ* plasmids, i.e. Glu ura⁻his⁻trp⁻ plates). The transformants are then scraped from the plates and frozen for storage. Aliquots are thawed and plated on a medium containing galactose and lacking leucine. This induces expression of the activation-tagged cDNA proteins and selects for transcription of the *LEU2* reporter. This two-step approach results in a uniform increase in the number of cells carrying each library plasmid; each cell transformed with the library is allowed to multiply to form a colony, and the colonies are harvested at approximately the same size. One advantage to this approach is that the number of transformed yeast is amplified before the synthesis of cDNA encoded proteins is induced. This ensures that yeast containing toxic or mildly toxic cDNA encoded proteins will not be depleted from the population.

The detailed procedures for the first step (transformation of the selection strain with the library DNA) are given in *Protocol 3* below. Following this, interactors are selected as described in *Protocol 4* (Section 4.2).

Protocol 3. Transforming the selection strain with library DNA [a]

Before transforming the strain with library DNA (which is usually fairly valuable), perform pilot transformations of the strain with the *TRP1* library vector. A transformation efficiency of at least 10^5 transformants/µg DNA should be the goal.

Equipment and reagents

- Liquid dropout media (Glu ura⁻his⁻; see *Table 2*)
- Dropout plates (Glu ura⁻his⁻trp⁻; Gal/Raf ura⁻his⁻trp⁻; see *Table 2*)
- Sterile water
- LiOAc/TE: 10 mM Tris–HCl pH 7.5, 1 mM EDTA, 100 mM lithium acetate (prepare this using filter sterilized stocks of 1 M lithium acetate pH 7.5, 1 M Tris–HCl pH 7.5, and 0.5 M EDTA)
- Library DNA
- Single-stranded salmon sperm DNA (prepared according to Schiestl and Gietz; ref. 47)
- DMSO

- 40% (w/v) PEG 4000 in LiOAc/TE: prepare this from filter sterilized 1 M lithium acetate pH 7.5, 1 M Tris–HCl pH 7.5, 0.5 M EDTA, and 50% (w/v) PEG 4000 (Sigma) in water
- Sterile TE: 10 mM Tris–HCl pH 7.5, 1 mM EDTA
- Glycerol solution: 65% (v/v) glycerol, 0.1 M MgSO₄, 25 mM Tris–HCl 7.4 (sterilize by autoclaving)
- Glass beads (4 mm diameter; Fisher Scientific) (sterilize the beads by autoclaving)
- Sterile 50 ml Falcon tubes
- Sterile 50 ml round-bottom polypropylene centrifuge tubes

A. Transformation

1. Grow yeast containing the bait and *lacZ* reporter plasmids in 400 ml of Glu ura⁻his⁻ medium at 30 °C, with shaking ($\sim$ 150 r.p.m.) to an OD_{600} of 1.0, corresponding to about 3×10^7 cells/ml. The doubling time in this medium is rather slow (2–4 h). For this reason it is sometimes convenient to start a smaller, 50 ml culture, grow it to $OD_{600} = 2.0$ or greater, and then dilute it into 400 ml. For high transformation efficiencies it is important to start the 400 ml culture at $OD_{600} = 0.2$ or less and allow it to grow to $OD_{600} = 1.0$ so that the cells are in log phase when harvested.

2. Centrifuge the culture at 2000 *g* for 5 min and pour off the supernatant. Wash the cells in 20 ml sterile water. These centrifugation steps and all following manipulations are carried out at 20–25 °C.

3. Resuspend the yeast in 5 ml of sterile LiOAc/TE. Pellet the cells again by centrifugation and pour off the supernatant.

4. Resuspend the yeast in 2 ml sterile LiOAc/TE; 50 µl of this suspension provides enough competent cells to be transformed with 1 µg of DNA.[b]

5. Aliquot 100 µl of the competent yeast into sterile microcentrifuge tubes. To each tube add 2 µg of library DNA and 60 µg of carrier (single-stranded salmon sperm DNA) in a total volume of 20 µl. The number of tubes to be used depends on the desired number of transformants and the expected transformation efficiency. If pilot transformations resulted in efficiencies of 10^5 transformants/µg DNA, each tube should yield 200 000 transformants, all of which can be plated on a single 24 cm × 24 cm plate (see part B).

6. Add DMSO to 10% (v/v). This improves transformation efficiency by three- to fivefold (48) (R. L. Finley and B. Cohen, unpublished data).

7. Add 600 µl of sterile 40% PEG 4000 in LiOAc/TE. Gently invert the tube several times to mix.

8. Incubate at 30 °C for 30 min. Agitation is not necessary.

9. Heat shock at 42 °C for 15 min and return to room temperature.

10. Determine the total number of transformants by removing 10 µl from each tube and making three dilutions (10-, 10^2-, and 10^3-fold) in sterile water. Plate 100 µl of each dilution on to 100 mm diameter Glu ura⁻his⁻trp⁻ plates and incubate at 30 °C. Calculate the total number of transformants from the number of colonies on these dilution plates.

At this point the transformation mixes can be plated to first select for all library transformants (as discussed in Section 4.1.1). This method is described below. Alternatively, the transformation mix can be used in a one-step selection for interactors (as discussed in Section 4.1.2); to perform this, proceed to *Protocol 4* (Section 4.2).

Protocol 3. *Continued*

B. *Selecting library transformants*

1. Plate each transformation mix (less than 1 ml) on to a single 24 cm ×
   24 cm Glu ura⁻his⁻trp⁻ plate. There is no need to spin the cells or
   remove the PEG. The medium in these plates should be at least 0.6 cm
   thick, level, and free of bubbles. To achieve an even distribution of
   cells, pour about 100 sterile glass beads (4 mm diameter) on to the
   plate with the cells. Gently roll the beads around the plate to distribute
   the transformation mix, then pour the beads off, or onto the next
   plate. Alternatively, distribute the transformation mix with a sterile
   bent glass rod. Both techniques work best when the surface of the
   plates is not too wet so that the medium absorbs the transformation
   mix. To achieve this moisture content, put newly solidified plates into
   a laminar flow hood with the lids ajar for about 2 h before plating.

2. Incubate the plates at 30 °C. Colonies should appear after about 24 h.
   Continue incubation until colonies are 1–2 mm in diameter, which
   should take a total of approximately two days.

C. *Harvesting the transformants*

1. Place the plates at 4 °C for 2–4 h to harden the agar.

2. Using the long edge of a sterile glass microscope slide (and sterile
   technique), scrape the yeast from the plate. Try not to scrape any agar
   as this will interfere with pipetting. Collect the yeast from the glass
   slide by wiping it on the lip of a sterile 50 ml Falcon tube.

3. Wash the cells twice with 2 or 3 vol. of sterile TE each time. It is best to
   pellet the cells each time in a sterile round-bottom polypropylene tube
   at 2000 *g* for 4 min so they may be easily resuspended. The pellet
   volume for 500 000 transformants will be about 8 ml.

4. Resuspend the cells thoroughly by swirling in one pellet volume of
   sterile glycerol solution. Mix well by vortexing on low speed. Freeze
   1 ml aliquots at − 70 °C.

5. Determine the plating efficiency by thawing an aliquot of library
   transformants and making serial dilutions in sterile water. Plate 100 µl
   of each dilution on to 100 mm diameter Gal/Raf ura⁻his⁻trp⁻ plates.
   Count the colonies that grow after two to three days at 30 °C.
   Represent the plating efficiency in colony-forming units (c.f.u.) per
   unit volume of frozen cells. The plating efficiency will be on the order
   of 10^8 c.f.u./100 µl.

---

[a] The transformation protocol given here is a variation of the high efficiency lithium acetate
method developed by Geitz *et al.* (36). Other high efficiency yeast transformation protocols, e.g.
electroporation (34, 35), may be substituted.

b Use of more than 1 μg DNA/50 μl of cells can result in introduction of multiple library plasmids into each yeast cell. This should be avoided because of the large amount of additional work that will be required to determine which of the library plasmids in a cell expresses the cDNA encoded protein that interacts with the bait.

4.1.2 One-step selection for interactors

A simpler, but so far inferior, alternative approach is to perform a one-step selection for yeast containing activation-tagged cDNA proteins that interact with the bait. This method involves plating the library transformation mix directly on plates lacking leucine (i.e. Gal/Raf ura⁻his⁻trp⁻leu⁻ plates) to select for expression of the *LEU2* reporter, without first selecting cells transformed with the library plasmid. Although it is easier than the two-step method, it suffers from some disadvantages. First, only a fraction of the yeast containing activation-tagged proteins that interact with the bait survive on the leu⁻ selection plates, possibly because some cells die during the time it takes to induce synthesis of the activation-tagged protein, the *LEU2* product, and leucine. This lower plating efficiency on the leu⁻ selection plates is particularly evident when the yeast contain an activation-tagged cDNA encoded protein that interacts only weakly with the bait. Secondly, the one-step approach may reduce the probability of isolating cDNAs that encode proteins that are somewhat toxic to yeast. These proteins will also cause a reduced plating efficiency on the leu⁻ plates. Although both of these disadvantages might be overcome by plating very large numbers of library transformants, obtaining such numbers would require exceptional yeast transformation efficiencies or huge amounts of library DNA. Furthermore, if expression of a given activation-tagged cDNA protein causes yeast to plate at reduced efficiency, a much larger number of yeast will need to be characterized in order to find it. For these reasons we prefer the two-step method (Section 4.1.1). However, we also present the best current version of the one-step method, which may eventually be improved to overcome the relative disadvantages described above. To do the one-step method follow *Protocol 3A* and then skip to *Protocol 4* (Section 4.2).

4.2 Isolating yeast with galactose-dependent Leu⁺ and lacZ⁺ phenotypes

Library transformants containing cDNAs that encode proteins that interact with the bait will exhibit galactose-dependent growth on media lacking leucine (Leu⁺), and galactose-dependent β-galactosidase activity (lacZ⁺). Isolation of these galactose-dependent Leu⁺ and lacZ⁺ yeast is accomplished in two steps:

(a) First, library transformants are plated on galactose medium lacking leucine (leu⁻) to select yeast that are Leu⁺.

(b) Secondly, the Leu$^+$ yeast are isolated, placed on a glucose master plate, and then replica plated to four new plates to test for *lacZ* expression and galactose dependence. These four plates include two leu$^-$ plates and two X-Gal plates: one leu$^-$ plate and one X-Gal plate contain galactose to induce cDNA expression (plus raffinose to enhance growth), while the other leu$^-$ plate and the other X-Gal plate contain glucose to repress cDNA expression. Yeast that grow on leu$^-$ galactose medium but not on leu$^-$ glucose medium, and that turn blue on galactose X-Gal plates but remain white on glucose X-Gal plates (i.e. those that are galactose-dependent Leu$^+$ and lacZ$^+$) are picked for further characterization.

In this procedure the yeast are grown on the glucose master plates to shut off cDNA expression before replica plating so that the galactose dependence of the Leu$^+$ and lacZ$^+$ phenotypes can be assessed. This avoids a problem that can arise when the Leu$^+$ yeast are replica plated from galactose to glucose plates, namely there may be sufficient mRNA and protein product from the activation-tagged cDNA protein to allow the yeast to grow on leu$^-$ glucose for several generations and turn blue on glucose X-Gal without further cDNA expression. This would mask the galactose-dependence of the Leu$^+$ and lacZ$^+$ phenotypes. The detailed procedures for selecting interactors using this approach are described in *Protocol 4*.

Protocol 4. Selecting interactors

If Trp$^+$ library transformants were selected and stored frozen (i.e. two-step method, Section 4.1.1), follow 'a' in the steps below. If library transformants were not selected and stored (i.e. one-step method, Section 4.1.2), follow 'b' in the steps below.

Equipment and reagents

- Liquid dropout media (Gal/Raf ura$^-$his$^-$trp$^-$; see *Table 2*)
- Dropout plates (Gal/Raf ura$^-$his$^-$trp$^-$leu$^-$, Glu ura$^-$his$^-$trp$^-$leu$^-$, Glu ura$^-$his$^-$trp$^-$; see *Table 2*)
- X-Gal plates (Glu ura$^-$his$^-$trp$^-$ X-Gal, Gal/Raf ura$^-$his$^-$trp$^-$ X-Gal; see *Table 2*)
- Applicator sticks or toothpicks sterilized by autoclaving

Method

1. Induce synthesis of activation-tagged cDNA encoded proteins as follows:

    (a) If library transformants were selected and frozen, thaw an aliquot of the cells and dilute tenfold into Gal/Raf ura$^-$his$^-$trp$^-$ liquid medium.

    (b) If performing the one-step method to select for interactors, dilute the transformation mix from *Protocol 3A*, step 9, tenfold into Gal ura$^-$his$^-$trp$^-$ media.

2. In each case, incubate at 30 °C with shaking for 4 h to induce the *GAL1* promoter. There is almost no increase in cell number during this time, and any increase can be neglected when calculating the number of colony-forming units or transformants to plate on to leu⁻ selection plates (step 4 below).

3. Pellet the cells by centrifugation at 2000 *g* for 4 min at 20–25 °C and resuspend in sterile water.

4. Plate on to Gal/Raf ura⁻his⁻trp⁻leu⁻ plates.

   (a) If library transformants were pre-selected and frozen, plate 10^6 c.f.u. (determined from the plating efficiency test in *Protocol 3C*, step 5) onto each 100 mm diameter plate.

   (b) If performing the one-step selection, plate an amount of the transformation mix that should contain 10^6 transformants (based on estimates from transformation efficiencies obtained in pilot transformation experiments) onto each 100 mm plate. Determine the actual number of library transformants plated by counting the colonies that grow when dilutions of the transformation mix are plated onto Gal/Raf ura⁻his⁻trp⁻ as described in *Protocol 3A*, step 10.

5. Incubate the selection plates at 30 °C. Colonies should appear in two to five days. To keep the plates from drying out after two days, it may be helpful to put them in plastic bags or containers, or put Parafilm around each plate. Generally, there will be more galactose-dependent Leu⁺ and lacZ⁺ yeast among the colonies that appear sooner, and fewer among the colonies that appear later.

6. Pick colonies with sterile toothpicks or applicator sticks and patch, or streak for single colonies, onto another Gal/Raf ura⁻his⁻trp⁻leu⁻ plate. Ideally the Leu⁺ yeast should be streaked for single colonies on leu⁻ galactose plates to isolate them from contaminating Leu⁻ yeast that were present when the Leu⁺ colony was forming. However, when there are large numbers of Leu⁺ colonies, it may be inconvenient to streak purify every one; in this case, growth of patches on a second selection plate will at least enrich for the Leu⁺ cells.

7. To show that the Leu⁺ phenotype is galactose-dependent, patch the Leu⁺ yeast on to Glu ura⁻his⁻trp⁻ master plates to turn off the *GAL1* promoter and stop expression of the activation-tagged cDNA protein. Grow at 30 °C for about 24 h.

8. Replica the master plates to the following four plates: 1. Glu ura⁻his⁻ trp⁻ X-Gal; 2. Gal/Raf ura⁻his⁻trp⁻ X-Gal; 3. Glu ura⁻his⁻trp⁻leu⁻; 4. Gal/Raf ura⁻his⁻trp⁻leu⁻. Incubate at 30 °C and examine the results after one, two, and three days.

Protocol 4. *Continued*

9. Pick only those yeast that are Leu$^+$ and lacZ$^+$ on galactose (for example, see Section 5, *Figure 5*). Further characterize these by isolating the library plasmid and determining the interaction specificity.

Three alternative procedures to the method described in *Protocol 4* have been used successfully:

(a) If a large number of Leu$^+$ yeast colonies appear on the initial leu$^-$ selection plates, those that are lacZ$^+$ can be quickly identified using a filter β-galactosidase assay. The filter assay, described in detail elsewhere (49), involves lifting a replica of the yeast from the colonies on the leu$^-$ selection plate with a nitrocellulose filter, lysing the yeast on the filter, and exposing the filter to buffer and X-Gal. The filter is then examined for blue spots corresponding to lacZ$^+$ yeast. Leu$^+$ yeast that correspond to those that are also lacZ$^+$ are then picked from the original leu$^-$ selection plate, put on to glucose master plates, and replica plated as described in *Protocol 4*, steps 7 and 8.

(b) The second alternative is to include X-Gal in the initial leu$^-$ selection plates and pick those that grow and turn blue. Again, the Leu$^+$ lacZ$^+$ yeast are picked, patched on to glucose master plates, and replica plated to the two sets of indicator plates. The disadvantage to this approach is that yeast grow less well on X-Gal plates because these plates contain a buffer to give them neutral pH; the reduced growth rate at this higher pH reduces the plating efficiency during the selection.

(c) Finally, the yeast colonies on the original leu$^-$ selection plate can be replica plated directly to X-Gal plates to determine which are lacZ$^+$ (A. Mendelsohn, personal communication). Only those that are Leu$^+$ and lacZ$^+$ are isolated and tested for galactose dependence.

Most activation-tagged proteins that interact strongly with the bait will render the yeast containing them galactose-dependent Leu$^+$ and lacZ$^+$. However, yeast containing activation-tagged cDNA proteins that interact only weakly with the bait may be galactose-dependent Leu$^+$, yet may appear light blue or even white on X-Gal plates. This is due to the different sensitivities of the Leu and lacZ phenotypes (see Section 2.2). Results from several interactor hunts have shown that there are usually more biologically relevant interactors in the class of yeast that are galactose-dependent Leu$^+$ and lacZ$^+$ than in yeast that are lacZ$^-$ (R. L. Finley, unpublished data). However, the class of yeast that are galactose-dependent Leu$^+$ but lacZ$^-$ may also contain biologically relevant interactors (R. L. Finley and R. Brent, in preparation). For this reason the decision whether or not to further characterize the latter class must be somewhat arbitrary.

5. Verifying specificity

A finding of galactose-dependent Leu$^+$ and lacZ$^+$ in a yeast isolate can be considered a demonstration that the reporter genes are activated due to expression of the activation-tagged cDNA encoded protein. However, it is important to determine that activation of the reporters is due to specific interaction of this protein with the bait, rather than to its non-specific interaction with LexA, with the promoters, or with some part of the transcription machinery. To verify that the cDNA encoded protein interacts specifically with the bait, library plasmids are rescued from the galactose-dependent Leu$^+$ lacZ$^+$ yeast and re-introduced into the original selection strain and into other strains containing different baits. Specific interactors confer the galactose-dependent Leu$^+$ and lacZ$^+$ phenotype to yeast containing the original bait, but not to yeast containing unrelated baits. To test this, master plates are made from the new transformants and replica plated on to four indicator plates as described in *Protocol 4*, steps 7 and 8 (see Section 5.2, *Figure 5*). Alternatively, the specificity can be determined using a mating assay (see Section 5.2, *Protocol 6*).

5.1 Rescue of library plasmids from yeast

Library plasmids are rescued from yeast by performing a quick yeast plasmid mini-prep and using the mini-prep DNA to transform *E. coli* (most yeast mini-prep protocols do not provide enough clean plasmid DNA for restriction analysis). If a large number of galactose-dependent Leu$^+$ lacZ$^+$ yeast are obtained, it is useful to reduce the number of library plasmids that need to be rescued by determining which ones contain identical cDNAs. This can be done by comparing restriction digests of PCR products containing the cDNA insert. In this procedure, yeast mini-prep DNA is used as template in PCR reactions with primers derived from sequences in the library plasmid flanking the cDNA insertion site. PCR products are then digested with one or two restriction enzymes that cut frequently. This procedure is described in *Protocol 5*.

Protocol 5. Isolating and classifying library plasmids[a]

Equipment and reagents

- TE: 10 mM Tris–HCl pH 7.5, 1 mM EDTA
- S buffer: 10 mM K$_2$H PO$_4$ pH 7.2, 10 mM EDTA, 50 mM 2-mercaptoethanol, 50 μg/ml zymolase
- Lysis solution: 25 mM Tris–HCl pH 7.5, 25 mM EDTA, 2.5% SDS
- 3 M potassium acetate pH 5.5
- *Taq* polymerase (5 U/μl; Boehringer-Mannheim Biochemicals)
- 10 × *Taq* polymerase buffer (Boehringer-Mannheim Biochemicals)
- dNTP mix (all four dNTPs at 2.5 mM each)

- 5′ primer (0.1 μg/μl) derived from the cDNA fusion moiety sequence (see *Table 3*)
- 3′ primer (0.1 μg/μl) derived from the *ADH1* terminator sequence in the library vector (see *Table 3*)
- *Alu*I and appropriate 10 × restriction enzyme buffer
- *Hae*III and appropriate 10 × restriction buffer
- 1.5% agarose gels, and other reagents and equipment for gel electrophoresis
- Thermal cycler for PCR reactions
- Electroporator

Protocol 5. *Continued*

A. *Preparation of yeast DNA mini-prep*[a]

1. Scrape a large mass of yeast from a plate and resuspend it in 1 ml of TE in a microcentrifuge tube (the OD_{600} of this suspension should be between two and five). Yeast from a fresh, two to three day-old plate work best. The yeast can also be obtained from a 1 ml overnight liquid culture.

2. Spin briefly to pellet the cells. Resuspend the yeast in 0.5 ml of S buffer.

3. Incubate at 37 °C for 30 min.

4. Add 0.1 ml lysis solution. Vortex to mix.

5. Incubate at 65 °C for 30 min.

6. Add 166 μl 3 M potassium acetate. Chill on ice for 10 min.

7. Spin in a microcentrifuge for 10 min. Pour the supernatant into a new tube.

8. Precipitate the DNA by adding 0.8 ml cold ethanol. Incubate on ice for 10 min, spin for 10 min, and pour off the supernatant.

9. Wash the pellet by centrifugation with 0.5 ml of 70% ethanol and dry the pellet.

10. Dissolve the pellet in 40 μl sterile water. Use 1–2 μl of this crude yeast mini-prep to transform *E. coli* by electroporation.[b]

B. *Classifying interactors*[b]

1. Set up a 0.5 ml tube for each yeast mini-prep containing:
   - 13 μl sterile water
   - 2 μl 10 × *Taq* polymerase buffer
   - 2 μl dNTP mix
   - 1 μl 5′ primer (0.1 μg/μl)
   - 1 μl 3′ primer (0.1 μg/μl)
   - 0.2 μl *Taq* polymerase (5 U/μl)

2. Add 1 μl of yeast mini-prep DNA to each tube.

3. Incubate in a thermal cycler for 25 cycles of 92 °C for 30 sec, 65 °C for 2 min, 75 °C for 30 sec.

4. Set up two tubes for each PCR reaction, one for *Alu*I digestion and one for *Hae*III digestion (these enzymes are recommended because they work well in the presence of the PCR reactants). Add 8 μl of the PCR reaction to each tube. Save the remainder of the PCR reaction for gel analysis of the full-length PCR product.

5. Add 1 μl of the appropriate 10 × restriction enzyme buffer to the 8 μl. Add 1 μl of *Alu*I to one tube and 1 μl of *Hae*III to the other tube. Incubate at 37 °C for 2 h.

6. Analyse *Alu*I digests, *Hae*III digests, and the uncut PCR products by electrophoresis on 1.5% agarose gels. It should be readily apparent from this analysis which cDNAs are identical. In some cases, only some of the restriction fragments will appear identical between two plasmids, suggesting that these plasmids contain different length cDNAs made from the same gene. Rescue library plasmids from those yeast mini-preps that give different restriction patterns.

[a] Several effective methods are available for isolating plasmids from yeast in amounts sufficient for *E. coli* transformation (e.g. the 'smash and grab' method of Hoffman and Winston ref. 50). The method described in *Protocol 5* is quick and yields plasmid DNA clean enough to transform *E. coli* efficiently and to work as a template for PCR.
[b] Before transforming *E. coli* with the crude yeast mini-preps it is often useful to determine which yeast mini-preps contain the same library plasmid so that fewer need to be characterized. This can be done by restriction analysis of PCR products.

Since the yeast contain three different plasmids, rescuing the library plasmid depends on distinguishing it from the bait and *lacZ* reporter plasmids. There are at least three ways to do this. In the most efficient one, yeast minipreps are used to transform a strain of *E. coli* that contains a mutation in the *trpC* gene. The inability of the *E. coli* to grow in the absence of tryptophan is complemented by the yeast *TRP1* gene on the library plasmid (51). *TrpC E. coli* transformed with the *TRP1* library plasmid are selected on minimal plates lacking tryptophan and containing ampicillin. In another method, yeast mini-preps are used to transform *E. coli*, and *E. coli* mini-preps from several individual transformants are analysed by restriction digestion to determine which mini-preps contain the library plasmid. Alternatively, the *E. coli* trans-

---

Table 3. Sequencing and PCR primers for pEG202 and pJG4-5

1. For sequencing inserts in the polylinker of pEG202, primers can be derived from LexA coding sequences. The primer shown below, LEX1, is derived from LexA coding sequences 40 bp upstream of the *Eco*RI site in the polylinker (18).[a]
2. For sequencing from the 5′ end of cDNA inserts in pJG4-5, BCO1 can be used. It is derived from the coding sequence for the B42 activation domain 70 bp upstream of the *Eco*RI site.
3. For sequencing from the 3′ end of the cDNA insert in pJG4-5, BCO2 can be used. It is derived from the sequence of the *ADH1* terminator approximately 40 bp downstream of the *Xho*I site.
4. BCO1 and BCO2 can be used for PCR amplification of the cDNA insert as described in *Protocol 5*.
   LEX1   5′ CGT CAG CAG AGC TTC ACC ATT G 3′
   BCO1   5′ CCA GCC TCT TGC TGA GTG GAG ATG 3′
   BCO2   5′ GAC AAG CCG ACA ACC TTG ATT GCA G 3′

[a] All eukaryotic LexA expression plasmids lack the *Eco*RI site found in the wild-type LexA coding region.

formations are plated on to LB amp plates containing X-Gal; transformants that contain the *lacZ* reporter plasmid form light blue colonies and are not picked. Restriction analysis on mini-preps from the white colonies is used to distinguish between the library and bait plasmids.

5.2 Demonstrating interactor specificity

To verify that the cDNA encoded protein interacts specifically with the bait, the rescued library plasmid DNA is re-introduced into the original selection strain and into other strains containing different baits. The procedure is described in *Protocol 6*.

Protocol 6. Determining specificity of interactors

Equipment and reagents

- Rescued library plasmid DNA
- Liquid dropout media (Glu ura⁻ his⁻; see Table 2)
- Dropout plates (Glu ura⁻ his⁻ trp⁻, Gal/Raf ura⁻ his⁻ trp⁻ leu⁻, Glu ura⁻ his⁻ trp⁻ leu⁻; see Table 2)

- X-Gal plates (Glu ura⁻ his⁻ trp⁻ X-Gal, Gal/Raf ura⁻ his⁻ trp⁻ X-Gal; see Table 2)
- Applicator sticks or toothpicks sterilized by autoclaving

Method

1. Use the rescued library plasmid DNA (from a mini-prep of transformed *E. coli*) to transform the original yeast strain containing the bait plasmid and the *lacZ* reporter plasmid. Additionally, use each library plasmid to transform one or more other strains containing different baits. Select transformants on Glu ura⁻ his⁻ trp⁻ plates. As a control, transform each different bait strain with the library vector without a cDNA insert.

2. Use sterile toothpicks or applicator sticks to pick three or four individual colonies from each transformation plate, and patch these to Glu ura⁻ his⁻ trp⁻ master plates. Patch control transformants (library plasmid with no cDNA) on to each plate for side-by-side comparison. Grow for one to two days at 30 °C.

3. Replica plate from the master plates to four plates: 1. Glu ura⁻ his⁻ trp⁻ X-Gal; 2. Gal/Raf ura⁻ his⁻ trp⁻ X-Gal: 3. Glu ura⁻ his⁻ trp⁻ leu⁻; 4. Gal/Raf ura⁻ his⁻ trp⁻ leu⁻.

4. Incubate plates at 30 °C and examine after one, two, and three days.

Figure 5 shows an example of the result obtained for three specific interactors.

6. Using a mating assay to verify specificity

We have recently developed a mating assay as an alternative way to test for interaction between a given activation-tagged protein and a panel of LexA

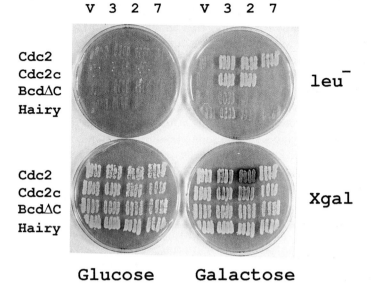

Figure 5. Specificity test. Specificity of *Drosophila* Cdc2 kinase interactors (Cdis; R. L. Finley and R. Brent, in preparation). Four replica plates were made from a glucose master plate (not shown) that contained EGY48 derivatives with different bait plasmids and library plasmids. All strains contained the *lacZ* reporter plasmid pJK103, and one of four bait plasmids directing the synthesis of LexA fused to *Drosophila* proteins: Cdc2 kinase (Cdc2), a Cdc2 kinase analogue (Cdc2c), a derivative of the homeodomain protein Bicoid (Bcd△C), or the helix–loop–helix protein Hairy (Hairy). Each of these bait strains was transformed with the library vector pJG4-5 (v), or with pJG4-5 that contained cDNAs that encode Cdc2 kinase interactors: Cdi3 (3), Cdi2 (2), and Cdi7 (7). Four individual colonies from each transformation were patched on to a glucose master plate which was then replica plated to two ura⁻his⁻trp⁻leu⁻plates (leu⁻), and two ura⁻his⁻trp⁻ X-Gal plates (Xgal). The plates on the *left* have glucose and the plates on the *right* have galactose and raffinose. Interaction is detected by growth of strains on the galactose leu⁻ plate; e.g. Cdc2 interacts with Cdi3, Cdi2, and Cdi7; Cdc2c interacts with Cdi3 and Cdi2. The strength of the interactions is suggested by the level of activation of *lacZ* as indicated on the X-Gal plate; e.g. the interaction between Cdc2 and Cdi2 activates *lacZ* strongly, the interaction between Cdc2 and Cdi7 activates *lacZ* very weakly, and the other interactions activate *lacZ* moderately.

fusion proteins (baits). This scheme takes advantage of the fact that haploid cells of the opposite mating type will fuse to form diploids when brought into contact with each other (52). In this mating assay, the activation-tagged protein is expressed in one yeast strain and the bait is expressed in a second strain of the opposite mating type. When the two strains are mixed on the same plate, they form diploids in which the bait and activation-tagged proteins have the opportunity to interact and activate the reporter genes. As before, interaction is measured as activation of the *LexAop-LEU2* and *LexAop-lacZ* reporters. This technique can be used to check the specificity of the cDNA encoded proteins isolated in the interaction trap to ensure that they interact with only the original bait and not with unrelated LexA fusions. It can also be used to examine interactions between a given set of activation-tagged proteins, such as those isolated in an interactor hunt, and a large panel of bait proteins without performing an unwieldy number of yeast transformations. Finally, it could be used to conduct interactor hunts; the library plasmids can be introduced in bulk into EGY48, transformants can be stored frozen, and then thawed and mated with a second strain that contains a bait plasmid. This can enable several separate interactor hunts with different baits to be done by performing a single large scale library transformation. The disadvantage of the mating assay is that the sensitivity of the reporters to activation by baits and activation-tagged proteins that interact with them is generally less in diploid cells relative to haploid cells. To minimize the effect of this difference, the most sensitive *LEU2* reporter is used.

As illustrated in *Figure 6*, the activation-tagged protein is expressed in EGY48 (mating type α) and the bait protein is expressed in a second strain, RFY206 (mating type *a*) which may also contain the *lacZ* reporter plasmid. The EGY48 derivatives are streaked on to plates lacking tryptophan to maintain selection for the library plasmid, and the RFY206 derivatives are streaked on to plates lacking uracil and histidine to maintain selection for the *URA3 lacZ* plasmid and the *HIS3* bait plasmid. The two strains are mated by applying them to the same replica velvet or filter and lifting their 'print' with a YPD plate. The YPD plate is incubated overnight at 30 °C to promote mating and then replica plated to the same indicator plates used in *Protocol 4*, steps 7

Figure 6. The mating assay for specificity. (a) A typical mating assay (R. L. Finley and R. Brent, in preparation). The his⁻ glucose plate (*top left*) contains seven RFY206 derivatives streaked horizontally. Each derivative contains a different *HIS3* bait plasmid. The trp⁻ glucose plate (*top right*) contains four EGY48 derivatives streaked vertically. Each derivative contains a different *TRP1* library plasmid. The RFY206 derivatives are MAT *a*, *HIS3*, trp1⁻, and Leu⁻. The EGY48 derivatives are MAT α *TRP1*, his3⁻, and Leu⁻. The two plates are pressed to the same replica velvet or filter and the replica is fitted with a YPD plate (*centre*). During overnight incubation at 30 °C the two strains grow. At the intersections on the YPD plate the two strains mate and form *His⁺ Trp⁺* diploids. The YPD plate is then replica plated to three plates: 1. a his⁻trp⁻ galactose plate (with raffinose), on which diploids grow, but neither haploid parent grows; 2. a his⁻trp⁻leu⁻ glucose plate, on which the activation-tagged cDNA encoded proteins are not expressed, *LEU2* is not transcribed, and no strains grow; 3. a his⁻trp⁻leu⁻ galactose plate (with raffinose), on which activation-tagged cDNA encoded proteins are expressed, interact with the baits, activate the *LEU2* gene, and allow growth. (b) Results of a mating assay. The horizontally streaked strains are RFY206 derivatives expressing LexA fusions to *Drosophila* Cdc2 kinase (Cdc2); the Cdc2 kinase analogue, Cdc2c (Cdc2c); a Bicoid derivative (BcdΔC); Hairy (Hairy); the budding yeast Cdc2 kinase homologue, Cdc28 (Cdc28), and two proteins isolated from a hunt for Cdc2c interactors (Cdi5 and Cdi11) (R. L. Finley and R. Brent, in preparation). The vertically streaked strains are EGY48 derivatives that contain the pJG4-5 library vector without a cDNA insert (V) or with Cdc2 kinase interactor cDNAs, CDo2 (2), CDi3, (3), and CDi7 (7).

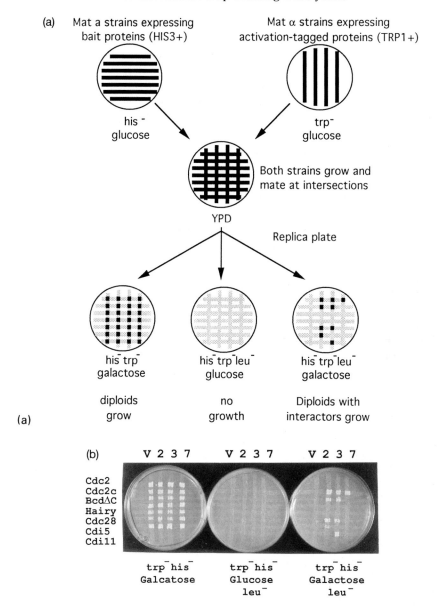

(a) Mat a strains expressing bait proteins (HIS3+)

Mat α strains expressing activation-tagged proteins (TRP1+)

his⁻ glucose

trp⁻ glucose

Both strains grow and mate at intersections

YPD

Replica plate

his⁻ trp⁻ galactose

his⁻ trp⁻ leu⁻ glucose

his⁻ trp⁻ leu⁻ galactose

diploids grow

no growth

Diploids with interactors grow

(a)

(b)

| V 2 3 7 | V 2 3 7 | V 2 3 7 |

Cdc2
Cdc2c
BcdΔC
Hairy
Cdc28
Cdi5
Cdi11

trp⁻ his⁻ Galcatose

trp⁻ his⁻ Glucose leu⁻

trp⁻ his⁻ Galactose leu⁻

and 8. In the example shown in *Figure 6*, the *lacZ* reporter is not used. The his⁻trp⁻ plates will contain only diploids: *TRP1* is provided by the strain that contains the library plasmid while *HIS3* is provided by the strain that contains the bait plasmid. Only the diploids that contain interacting pairs of activation-tagged protein and bait will grow on the plates lacking leucine but containing galactose.

The detailed procedures for the mating assay for specificity are described in *Protocol 7*.

Protocol 7. Mating assay

A. *Equipment and reagents*

- Rescued library plasmid DNA
- Liquid YPD medium (see *Table 2*)
- Liquid dropout media (Glu ura⁺; see *Table 2*)
- YPD plates (see *Table 2*)
- Dropout plates (Glu trp⁻, Glu ura⁻his⁻, Glu ura⁻his⁻trp⁻, Gal/Raf ura⁻his⁻trp⁻leu, Glu ura⁻his⁻trp⁻leu⁻; see *Table 2*)

- X-Gal plates (Glu ura⁻his⁻trp⁻ X-Gal, Gal/Raf ura⁻his⁻trp⁻ X-Gal; see *Table 2*)
- Applicator sticks or toothpicks sterilized by autoclaving
- Replica plating apparatus and sterile velvets or filters

Method

1. Introduce *TRP1* library plasmids into yeast strain EGY48 and select transformants on Glu trp⁻ plates. As a control, transform EGY48 with a library plasmid that has no cDNA insert.

2. Introduce *HIS3* bait plasmids into yeast strain RFY206 (or other *ura3 his3 trp1 leu2* MAT**a** strain) along with a *URA3 lacZ* reporter and select transformants on Glu ura⁻his⁻ plates.[a]

3. Use sterile toothpicks or applicator sticks to streak individual EGY48 transformants on to standard 100 mm Glu trp⁻ plates in parallel lines, six or seven to a plate (see *Figure 6*). Include at least one streak of the transformants with the control plasmid (no cDNA). Likewise, streak individual RFY206 transformants on to Glu ura⁻his⁻ plates in parallel lines, six or seven to a plate. Incubate at 30 °C until heavy growth. The lines of yeast should be at least 2 mm wide.

4. Onto the same replica filter or velvet, print the EGY48 derivatives and the RFY206 derivatives so that the streaks from the two plates are perpendicular to each other.

5. Lift the print of the two strains from the filter or velvet with a YPD plate. Incubate the YPD plate at 30 °C overnight. Diploids will form where the two strains intersect. One strain may grow more rapidly than the other during this time but this does not hinder formation of diploids in the intersections.

6. Replica from the YPD plate to the following plates: 1. Glu ura⁻his⁻trp⁻ X-Gal; 2. Gal/Raf ura⁻his⁻trp⁻ X-Gal; 3. Glu ura⁻his⁻trp⁻leu⁻; 4. Gal/Raf ura⁻his⁻trp⁻leu⁻. Incubate at 30 °C and examine after one, two, and three days. Only diploids will grow on the X-Gal plates and only interactors will grow on galactose plates lacking leucine (for example, see *Figure 6*).

[a] Note: in the example shown in *Figure 6* the *URA3 lacZ* reporter is not used.

---

7. Expected results

The two most critical parameters that determine whether an interactor hunt will succeed are the quality of the library and the nature of the bait. If the library contains cDNAs that encode proteins that interact with the bait in the interaction trap, the bait becomes the most critical parameter. The bait can affect the outcome of an interactor hunt in two ways:

(a) First, the number of non-specific cDNA encoded interactors obtained appears to depend on the bait. For example, use of some baits can result in isolation of activation-tagged cDNA proteins that seem to interact with many other LexA fusions when the specificity determination is performed. It is unlikely that these cDNA encoded proteins interact with the LexA moiety or with the reporters or transcription machinery directly because if they did, they would be expected to arise in all interactor hunts. Rather, these cDNA encoded proteins may simply be 'sticky', for example, because they contain patches of hydrophobic or charged residues that interact with corresponding regions on many baits.

(b) Second, the number of spurious Leu^+ yeast that arise independently of the library plasmid differs from one bait to another. For example, when some selection strains are transformed with library DNA and plated on to media lacking leucine, hundreds of galactose-independent Leu^+ colonies grow. This phenomenon is observed more frequently with baits that activate a low level of transcription than with transcriptionally inert baits, perhaps because a low level of activation is readily enhanced in a fraction of cells in a population. For example, a bait may gain the ability to activate the reporters in some cells if there is an increase in the copy number of the bait plasmid, which could result from natural variation in plasmid copy number or from a mutation in a yeast gene that affects plasmid copy number. This problem can be best addressed by reducing the activation potential of a bait as described in Section 4.2.

It is worth noting that, in our experience, these problems with baits are usually surmountable, and that most proteins can eventually be used successfully in interactor hunts.

Acknowledgements

We are very grateful to the numerous current and past laboratory members, and to the many laboratory visitors whose efforts contributed to the development of this interaction technology. We thank Todd Gulick, Pierre Colas, and Andrew Mendelsohn for critical readings of the manuscript. R. L. F. was supported by a postdoctoral fellowship from the NIH. R. B. was supported by the Pew Scholar's program. Work was supported by the HFSP.

References

1. Keegan, L., Gill, G., and Ptashne, M. (1986). *Science*, **231**, 699.
2. Hope, I. A. and Struhl, K. (1986). *Cell*, **46**, 885.
3. Brent, R. and Ptashne, M. (1985). *Cell*, **43**, 729.
4. Ma, J. and Ptashne, M. (1987). *Cell*, **51**, 113.
5. Sadowski, I., Ma, J., Triezenberg, S., and Ptashne, M. (1988). *Nature*, **335**, 563.
6. Chasman, D. I., Leatherwood, J., Carey, M., Ptashne, M., and Kornberg, R. D. (1989). *Mol. Cell. Biol.*, **9**, 4746.
7. Ma, J. and Ptashne, M. (1988). *Cell*, **55**, 443.
8. Fields, S. and Song, O. (1989). *Nature*, **340**, 245.
9. Chien, C.-T., Bartel, P. L., Sternglanz, R., and Fields, S. (1991). *Proc. Natl. Acad. Sci. USA*, **88**, 9578.
10. Durfee, T., Becherer, K., Chen, P.-L., Yeh, S.-H., Yang, Y., Kilburn, A. E., Lee, W.-H., and Elledge, S. J. (1993). *Genes Dev.*, **7**, 555.
11. Gyuris, J., Golemis, E., Chertkov, H., and Brent, R. (1993). *Cell*, **75**, 791.
12. Vojtek, A. B., Hollenberg, S. M., and Cooper, J. A. (1993). *Cell*, **74**, 205.
13. Dalton, S. and Treisman, R. (1992). *Cell*, **68**, 597.
14. Silver, P. A., Keegan, L., and Ptashne, M. (1984). *Proc. Natl. Acad. Sci. USA*, **81**, 595.
15. Bram, R. J. and Kornberg, R. D. (1985). *Proc. Natl. Acad. Sci. USA*, **82**, 43.
16. Giniger, E., Varnum, S. M., and Ptashne, M. (1985). *Cell*, **40**, 767.
17. Silver, P. A., Brent, R., and Ptashne, M. (1986). *Mol. Cell. Biol.*, **6**, 4763.
18. Brent, R. and Ptashne, M. (1984). *Nature*, **312**, 612.
19. Yocum, R. R., Hanley, S., West, R. J., and Ptashne, M. (1984). *Mol. Cell. Biol.*, **4**, 1985.
20. Guarente, L. and Ptashne, M. (1981). *Proc. Natl. Acad. Sci. USA*, **78**, 2199.
21. Guthrie, C. and Fink, G. R. (ed.) (1991). *Guide to yeast genetics and molecular biology. Methods in enzymology*, Vol. 194. Academic Press, Inc., Boston.
22. Ausubel, F. M., Brent, R., Kingston, R. E., Moore, D. D., Seidman, J. G., and Struhl, K. (ed.) (1992). *Current protocols in molecular biology*. (Chapter 13). Greene and Wiley-Interscience, New York.
23. Rothstein, R. (1985). In *DNA cloning: a practical approach* (ed. D. M. Glover), Vol. II, pp. 45–66. IRL Press, Oxford.
24. Golemis, E. and Brent, R. (1992). *Mol. Cell. Biol.*, **12**, 3006.
25. Zervos, A. S., Gyuris, J., and Brent, R. (1993). *Cell*, **72**, 223.
26. Lech, K., Anderson, K., and Brent, R. (1988). *Cell*, **52**, 179.
27. Samson, M.-L., Jakson-Grusby, L., and Brent, R. (1989). *Cell*, **57**, 1045.
28. Hanes, S. and Brent, R. (1989). *Cell*, **57**, 1275.
29. Kamens, J., Richardson, P., Mosialos, G., Brent, R., and Gilmore, T. (1990). *Mol. Cell. Biol.*, **10**, 2840.
30. Ebina, B. A., Takahara, Y., Kishi, F., Nakazawa, A., and Brent, R. (1983). *J. Biol. Chem.*, **258**, 13258.
31. Gill, G. and Ptashne, M. (1987). *Cell*, **51**, 121.
32. Broach, J. R. (1981). In *The molecular biology of the yeast Saccharomyces: life cycle and inheritance* (ed. J. N. Strathern, E. W. Jones, and J. R. Broach), pp. 445–70. Cold Sping Harbor Laboratory, Cold Spring Harbor, NY.
33. West R. W. Jr., Yocum, R. R., and Ptashne, M. (1984). *Mol. Cell. Biol.*, **4**, 2467.

34. Manivasakam, P. and Schiestl, R. H. (1993). *Nucleic Acids Res.*, **21**, 4414.
35. Becker, D. M. and Guarente, L. (1991). In *Methods in enzymology* (ed. C. Guthrie and G. R. Fink), Vol. 194, pp. 182–7. Academic Press, Inc., Boston.
36. Gietz, D., St. Jean, A., Woods, R. A., and Schiestl, R. H. (1992). *Nucleic Acids Res.*, **20**, 1425.
37. Brent, R. (1985). *Cell*, **42**, 3.
38. Finley, R. L. Jr., Chen, S., Ma, J., Byrne, P., and West, R. W. Jr. (1990). *Mol. Cell. Biol.*, **10**, 5663.
39. Miller, J. (1972). *Experiments in molecular genetics*, Cold Spring Harbor Laboratory, Cold Spring Harbor, NY.
40. Finley, R. L. Jr. and West, R. W. Jr (1989). *Mol. Cell. Biol.*, **9**, 4282.
41. Laemmli, U. K. (1970). *Nature*, **227**, 680.
42. Harlow, E. and Lane, D. (1988). *Antibodies: a laboratory manual.* Cold Spring Harbor Laboratory, Cold Spring Harbor, NY.
43. Gill, G. and Ptashne, M. (1988). *Nature*, **334**, 721.
44. Berger, S. L., Pina, B., Silverman, N., Marcus, G. A., Agapite, J., Regier, J. L., Triezenberg, S. J., and Guarente, L. (1992). *Cell*, **70**, 251.
45. Gubler, U. and Hoffman, B. J. (1983). *Gene*, **25**, 263.
46. Huse, W. D. and Hansen, C. (1988). *Stratagene strategies*, **1**, 1.
47. Schiestl, R. H. and Gietz, R. D. (1989). *Curr. Genet.*, **16**, 339.
48. Hill, J., Ian, K. A., and Griffiths, D. E. (1992). *Nucleic Acids Res.*, **19**, 5791.
49. Breeden, L. and Nasmyth, K. (1985). *Cold Spring Harbor Symposia on quant. biol.* **50**, 643. Cold Spring Harbor, NY.
50. Hoffman, C.S. and Winston, F. (1987). *Gene*, **57**, 267.
51. Struhl, K., Stinchomb, D. T., Scherer, S., and Davis, R. W. (1979). *Proc. Natl. Acad. Sci. USA*, **76**, 1035.
52. Herskowitz, I. (1988). *Microbiol. Rev.*, **52**, 536.
53. Kalderon, D., Roberts, B. L., Richardson, W. D., and Smith. A. E. (1984). *Cell*, **39**, 499.
54. Green, N., Alexander, H., Olson, A., Alexander, S., Shinick, T. M., Sutcliffe, J. G., and Lerner, R. A. (1982). *Cell*, **28**, 477.

7

The baculovirus expression system

GUNVANTI PATEL and NICHOLAS C. JONES

Detailed biochemical and structural analysis of proteins encoded by cloned mammalian genes requires efficient expression systems for their high level expression and purification. Although very many proteins have been success-fully over-produced in bacteria, there are certain inherent disadvantages which often makes this approach unsuitable. Foreign proteins over-expressed in bacteria often accumulate as insoluble occlusion bodies which require harsh denaturing agents to purify the recombinant protein. In addition, pro-teins produced in bacteria are not post-translationally modified. Since in many cases protein modification is crucial to activity, their over-production and purification from bacteria will be of limited use in functional studies. This chapter describes the helper-independent baculovirus expression system which offers a convenient and efficient method of over-producing post-translationally modified proteins in a eukaryotic cell and may therefore be the method of choice under many circumstances. This system takes advantage of unique features in the life cycle of baculoviruses, including the production of two classes of viral progeny, and the presence of the strong polyhedrin promoter which is non-essential both for viral replication and the production of infectious extracellular virus. We describe the key features of this ex-pression system including the latest modifications which will facilitate the iso-lation of recombinant baculoviruses expressing foreign proteins. Descriptions of general DNA manipulations, protein analysis, tissue culture procedures, and handling of baculoviruses can be found in other laboratory manuals (1–5).

1. Baculoviruses

Baculoviruses constitute one of the largest and most diverse groups of patho-genic viruses. Baculovirus disease of insects was described in 1527 as the 'jaundice disease' of the silkworm *Bombyx mori*. In the 1940s the viral nature of the disease was demonstrated and the structure and physiology of the viruses were investigated in greater detail. Baculovirus infections are observed mostly in insect species (over 600 species of insects have been shown to be susceptible to different members of the viral family), although baculo-viruses pathogenic for crustacea have also been reported. Importantly, they are not known to have any non-arthropod hosts and are therefore considered safe for use in the laboratory. The large rod shaped enveloped virion contains a covalently closed circular double-stranded DNA genome of 88–200 kb (6) with the baculovirus nucleocapsids generally being 36–50 nm in diameter and 200–400 nm in length. The size of the viral genome determines the length of the capsids which can readily extend to accommodate large inserts within the genome.

The family Baculoviridae is divided into three subgroups: nuclear poly-hedrosis viruses (NPV), granulosis viruses, and non-occluded-type viruses. NPVs can have a single virion (SNPV) or multiple virions (MNPV) occluded within nuclear protein crystals known as polyhedra. Because of the relative ease of handling in tissue culture, the MNPV from the Lepidopteran alfalfa looper, *Autographa californica* (AcMNPV), which has a genome of 128 kb, has been the most intensively studied baculovirus at the molecular level (for reviews see refs 3, 5, 7, 8). Both AcMNPV and the virus isolated from the silkworm *Bombyx mori*, BmNPV, have been used as efficient expression vectors, with the latter virus used for producing recombinant protein in the silkworm larvae (9).

2. The baculovirus infectious cycle

The life cycle of the baculovirus AcMNPV is shown in *Figure 1*. The ability to propagate AcMNPV *in vitro* has led to a detailed understanding of the kinetics of viral gene expression and the molecular events that occur during an infectious cycle. The cell lines used for these studies, namely IPLB-SF-21 and its derivatives (e.g. Sf9 cells), derive from the fall armyworm *Spodoptera frugiperda*. The availability of such cell lines that can be easily cultured and infected has greatly facilitated not only the study of AcMNPV replication but also the development of the baculovirus expression system.

During infection, AcMNPV viral genes are expressed temporally in three phases: an early phase, a late phase, and a very late or occlusion-specific phase (10). Replication is biphasic, in that two structurally different but genetically identical virion phenotypes are produced during the normal life cycle. Following the infection by virus, a number of early genes that are distributed throughout the AcMNPV genome are expressed prior to the onset of DNA replication. Such early gene expression is essential in order for efficient viral replication to take place. This phase is followed by a late phase of viral gene expression which extends from approximately 6 hours post-infection (p.i.) through to 18 hours p.i. During this phase the virus undergoes a primary round of replication (11). Cytopathic effects observed during this phase include an enlarged nucleus and virogenic stroma (11). At about 12 hours p.i., the first progeny virus are released by budding through the plasma

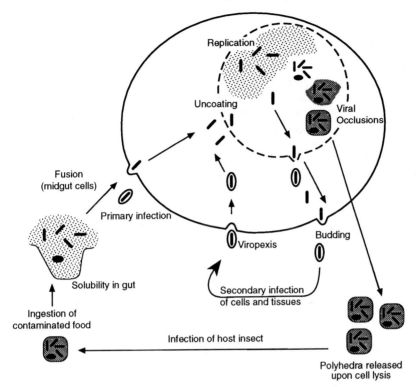

Figure 1. A schematic diagram representing the biphasic life cycle of a typical baculo-virus. The polyhedra ingested by larvae dissolve in the gut releasing the infectious virus. In the infected gut cell the viral DNA undergoes replication and extracellular virus (ECV) are released 10–12 hours post-infection by budding. These ECV can infect other cells by viropexis, thus spreading the infection. Late in infection, 20–24 hours post-infection, the viral nucleocapsids accumulate in the nucleus forming viral occlusions (polyhedra). (Adapted from ref. 2)

membrane. These are the non-occluded extracellular virus particles (ECV) or budded virus (BV). In the third phase of the infectious cycle, a set of specific genes, sometimes referred to as the very late or occlusion-specific genes, are expressed. This is accompanied by *de novo* membrane proliferation within the nucleus resulting in the envelopment of viral nucleocapsids. Bundles of such enveloped nucleocapsids accumulate in the nucleus and then become encapsulated within the matrix of crystalline occlusion bodies termed poly-hedra (12). By 70 hours p.i., wild-type AcMNPV produces an average of 70 occlusions per nucleus with each occlusion being approximately 2 μm in diameter, these being visible under the light microscope. Polyhedrin protein, with a molecular weight of 29 kd, constitutes the major structural component of the viral occlusions (13). In tissue culture cells, such as the Sf9 cell line, polyhedrin accumulates during AcMNPV infection to very high levels, accounting for as much as 50% of the total 'stainable' protein of the infected cell detected by SDS–PAGE (2). The expression of at least one other late gene, encoding the p10 protein, is regulated in a manner similar to that of polyhedrin (14). The p10 gene product appears to be involved in the formation of fibrous material observed in the occlusion-specific phase. The function of this material is unclear, but may be involved in processes leading to the disintegration of infected larvae.

Horizontal transmission of the virus in nature is dependent upon the production of occluded virus (OV). The OV is commonly found on plant surfaces and in the soil left by dead and decomposing tissues of infected larvae where it may persist for many years. Ingestion by new insect larvae is followed by solubilization of the occlusion bodies in the insect midgut resulting in the release of highly infectious virus. This mode of transmission is unimportant for growth *in vitro* and accordingly deletion or substitution of the polyhedrin gene is permissible and has no effect on the production of infectious ECV. The helper-independent baculovirus expression system is based on replacing this highly expressed viral polyhedrin gene with a recombinant gene consisting of the polyhedrin promoter fused upstream from the desired foreign coding sequence and subsequent transfer of this recombinant gene into the baculovirus genome by homologous recombination *in vivo* (3, 5, 15). The ensuing recombinant virus is fully infectious and at late times of infection the activated polyhedrin promoter directs efficient transcription of the foreign gene.

Although the polyhedrin promoter has been predominantly used for the expression of foreign genes, several workers have also constructed vectors that utilize the p10 promoter, another highly expressed very late promoter (3, 5). Additionally, some workers are investigating the use of promoters which are expressed earlier in the infection process thus allowing a longer time for the expression of the foreign protein before complete shut-off of host protein synthesis and pathways involved in post-translational modifications.

3. Advantages offered by the baculovirus expression system

The baculovirus insect cell system has been a popular choice for expression of foreign genes for a number of reasons:

(a) Expression of foreign genes in the eukaryotic insect cells provides a means for over-expressing soluble protein that is post-translationally modified.

(b) Baculoviruses are rod shaped viruses and are in theory able to incorporate relatively large segments of additional foreign sequences without much effect on replication efficiency or packaging of DNA within the nucleocapsid. The largest inserted fragment reported so far is approximately 15 kb (3).

(c) Recombinant baculoviruses are considered safe, since they are unable to replicate or integrate their genes in non-vertebrate cells. The expression system itself provides a level of containment since expression of a foreign gene in place of the polyhedrin gene limits the ability of the virus to survive under normal environmental conditions.

(d) The very late genes, such as polyhedrin and p10, are non-essential for normal replication and production of infectious extracellular virus, thus allowing the exploitation of these strong promoters for the expression of foreign genes after the shut down of host cell synthesis. In fact, using 10^9 infected cells, as much as 1–100 μg of non-baculovirus protein can be expressed under the direction of the polyhedrin promoter. Very late expression may also be particularly advantageous if the protein under investigation is toxic or deleterious to the host cell function. The negative aspect of this feature is that the protein expressed may not be fully modified since most of the host cell functions are declining or drastically reduced during the very late stage of infection.

(e) The baculovirus expression system is relatively simple to use and can be scaled up to larger volumes for the use in fermenter systems.

(f) Two or more foreign genes can be expressed simultaneously using the baculovirus system, thus allowing the study of complex formation between their protein products.

4. Possible disadvantages of the baculovirus expression system

(a) Since the consequence of infection is cell death, it is not possible for relevant protein production to be maintained in continuous culture.

(b) Although the baculovirus expression system provides a suitable environment for the production of post-translationally modified protein, in some cases modification, especially glycosylation (see below), of the recombinant protein is not identical to that found in mammalian cells.

However for most purposes, the advantages far outweigh the disadvantages of choosing baculoviruses to express foreign proteins. Furthermore, future modifications of the system are likely to be directed towards reducing these limitations, particularly for extending the capacity of the system for correct post-translation modifications.

5. Post-translational modification of proteins expressed in the baculovirus system

A wide-range of genes, too numerous to list, have been successfully expressed using the baculovirus expression system (see refs. 3, 5). Characterization of the recombinant proteins produced in baculovirus infected Sf9 cells has revealed the majority of them to be biologically active and to undergo post-translational modifications very similar to those of the authentic proteins expressed in their normal eukaryotic environment. Some of the post-translational modifications observed in insect cells include phosphorylation, glycosylation, acylation and proteolytic processing. In addition, the subcellular localization of expressed proteins is generally as expected and the system has also proved to be amenable to the study of tertiary and quarternary structure formation.

5.1 Phosphorylation

Several phosphoproteins including cellular transcription factors (16–18), cell cycle regulated genes (19, 20), viral proteins (21, 22), cellular receptors (23, 24) and enzymes (25) have been expressed successfully in insect cells and reported to be phosphorylated. In some cases the ensuring phosphorylation has been characterized in detail. Whilst it is clear that many phosphorylation events do occur, it is also clear that others occur inefficiently if at all. There could be many reasons for this, including low levels of the relevant endogenous kinase, a decrease in endogenous kinase levels at late times of infection when synthesis of the foreign protein is taking place or, under circumstances where expression of the foreign protein is particularly efficient, an excess of substrate overwhelming the capabilities of the kinase in question.

This heterogeneity in phosphorylation efficiency is exemplified by studies on the adenovirus E1A protein (21) and the SV40 T antigen protein (26). In both these cases the proteins are normally phosphorylated on multiple serine and threonine residues. Phosphorylation of these proteins does take place when they are over-expressed in insect cells and in both cases the phosphorylation patterns were found qualitatively to be remarkably similar to the pattern seen following expression in mammalian cells. However, clear differences were apparent in the efficiency of particular phosphorylations as demonstrated by the intensity of different species upon separation by two-dimensional electrophoresis or the extent of phosphate incorporation into different phosphopeptides. Depending on the aims of the experiment, these differences do not necessarily matter; in the case of both E1A and SV40 T antigen, the proteins that were expressed and isolated were found to be biologically active.

Phosphorylation of proteins on tyrosine residues following over-expression in the baculovirus system has also been reported. However the cases that showed the greatest degree of efficiency were those where the phosphorylation is due to autophosphorylation. For example, the human EGF receptor was found to be autophosphorylated on C-terminal tyrosine residues (23) and pp60[c-src] was autophosphorylated on Tyr 416 (22). Phosphorylation on tyrosine residues due to endogenous tyrosine kinases, such as phosphorylation of pp60[c-src] on Tyr 527 (22) and the phosphorylation of glycogen synthase kinase 3 α and β proteins (27), is generally found to be much less efficient.

In some cases the efficiency of phosphorylation on a particular site can be boosted by co-infection with a virus that can express the relevant kinase. Additionally, such co-infections can be exploited to characterize a particular phosphorylation such as to identify the kinase involved or study the functional consequence of a phosphorylation event. A case in point concerns the synthesis of pRB protein in insect cells which can be hyperphosphorylated by co-infection with viruses that express cyclin D and the cdk4 kinase (28). As a result, the hyperphosphorylated protein was shown to be no longer capable of forming stable ternary complexes with cyclins D2 and D3 or with the transcription factor E2F.

5.2 Glycosylation

Many glycoproteins have been synthesized in insect cells using the baculovirus expression system (5). In the majority of cases the glycosylation that occurs is N-linked, initiated by the addition of oligosaccharides to asparagine residues located in the recognition sequence Asn–X–Ser/Thr. In mammalian cells this core oligosaccharide is further modified by addition of fucose, galactose, and sialic acid residues to give complex branched oligosaccharide chains. Generally speaking, detailed analysis has revealed that in most cases glycosylation of proteins over-expressed in insect cells takes place on the Asn residues that are found to be glycosylated in mammalian cells, but that subsequent formation of complex structures does not occur. This is presumably due to lack of the appropriate transferase enzymes. These findings are not particularly surprising since it had been shown using viral systems that such complex glycosylation was deficient in insect cells.

In addition to N-linked glycosylation, there has been a report of successful O-linked glycosylation on a pseudorabies protein (gp50) following its expression in insect cells (29).

5.3 Fatty acid acylation

Radiolabelling with [3H]palmitic acid and [3H]myristic acid of infected insect cells has demonstrated two types of fatty acid acylation, palmitoylation and myristoylation. The baculovirus-expressed SV40 T antigen (30), Ha-ras p21 (31), and human transferrin receptor (32) have all been shown to be labelled with [3H]palmitic acid, while myristoylation has been shown to occur in the baculovirus-expressed hepatitis B surface antigen (33).

5.4 Proteolytic processing

Enzymes required for proteolytic cleavage of recombinant proteins appear to be present and functional in insect cells. Signal peptides are correctly processed from many membrane-bound or secreted proteins such as human α-interferon (34), human interleukin-2 (35), and human tissue plasminogen activator (36). Proteolytic processing events other than the cleavage of signal peptides have also been demonstrated to occur. For example, the baculovirus-expressed fowl plague virus HA protein was found to be functional indicating that correct processing had taken place, since activity is dependent on the cleavage of HA to HA1 and HA2 subunits (37). Proteolytic processing does not always occur, however, as demonstrated by the failure to process the HA protein of an influenza virus (38).

6. Construction of recombinant viruses

Owing to its large size, it is not possible to manipulate readily the genome of AcMNPV and insert into it foreign DNA by conventional techniques such as the use of restriction endonuclease cleavage sites. Therefore, the method that has been used relies upon allelic replacement involving cell-mediated homologous recombination and heterologous gene insertion. This involves the co-transfection into permissive cells of wild-type viral DNA together with an appropriate transfer vector that contains the foreign gene to be expressed downstream of the polyhedrin promoter. This chimeric gene is flanked by viral sequences that direct recombination resulting in the generation of recombinant viruses that contain the foreign gene in place of the polyhedrin coding sequences.

The frequency of recombination generally obtained is typically 0.1–1%. As such, the recombinant virus has to be identified over a very significant background of parental virus. Several methods have been used to identify the recombinant viruses including:

(a) The identification by microscopy of viral plaques that consist of polyhedra-negative (recombinant virus) infected cells (2).

(b) The identification of recombinant plaques by plaque hybridization to probes specific for the foreign DNA followed by visual screening (23, 39).

(c) Antibody screening of plaques (40).

For each of these methods, multiple plaque assays are generally required before pure recombinant virus is obtained.

Until recently, the time-consuming isolation of the recombinant baculovirus had been the main cause of frustration and the main disincentive to the use of this powerful expression system. More recently, however, several modifications of procedure have been developed that either result in easier identification of the recombinant virus or an increase in the proportion of progeny virus that are recombinant.

6.1 *LacZ*-based systems

The first significant modification used the *lacZ* gene as a marker to facilitate the identification of recombinant plaques. This gene can be used in three different ways:

(a) By inserting the *lacZ* gene in the parent virus in place of the polyhedrin gene and then identifying white (*lacZ⁻* recombinant), occ⁻ plaques over a background of blue, occ⁻ plaques (41). *LacZ⁺* plaques appear blue in the presence of the β-galactosidase substrate, X-Gal. In practice this method sometimes proved to be difficult, since the blue colour from the background of parent virus plaques could obscure the identification of recombinant plaques.

(b) By inserting the *lacZ* gene at a non-target site some distance from the polyhedrin gene. The recombinant virus like the parent produces blue plaques to aid their visualization (42). The recombinants carry a foreign gene in the place of the polyhedrin gene and produce blue occ⁻ plaques as opposed to blue occ⁺ plaques formed by the parent virus. The advantage of using such parent virus is that, because of the blue-plaque phenotype, the occ⁻ plaques should not be overlooked during the examination of the cell monolayer.

(c) By inserting the *lacZ* gene into the transfer vector itself [e.g. pBlueBac2 (43); pAcDZ1 (44); see Section 7.4]. The *lacZ* gene is thus transferred into the virus by recombination together with the polyhedrin promoter driven foreign gene so that the recombinant, but not the parent, plaques are blue in the presence of the appropriate β-galactosidase substrate.

6.2 Linearization of genomic DNA

A second significant modification resulted in an increase in the frequency of recombination as a result of linearizing the parent genome at a suitable target site. Linearized baculovirus DNA is generally non-infectious but is still re-combinogenic if homologous sequences (contained in the transfer vector) are also introduced into the cell. If these homologous sequences span the free ends of the linearized viral genome, a double cross-over event would result in a circular, infectious genome harbouring the foreign gene of interest. The observation that wild-type AcMNPV DNA does not contain a restriction site for the endonuclease *Bsu*361 was used by Kitts *et al.* (45) to construct a recombinant baculovirus that was engineered to contain a single recognition site for the *Bsu*361 (5′CCYNAGG3′) enzyme upstream of the polyhedrin promoter. The modification does not effect any of the functions of the virus. The use of the *Bsu*361 linearized genomic DNA in co-transfection protocols rather than circular genomic DNA resulted in a significant increase in the frequency of recombination (45).

This general method for producing recombinants has now evolved one step further by the construction of a parent virus engineered to contain a second *Bsu*361 restriction site (46). This second site is located in an essential gene (the 1629 ORF) which is located downstream of the polyhedrin gene. Therefore, restriction of such parental DNA with *Bsu*361 results in a large linear fragment that lacks certain essential genomic sequences as well as the non-essential polyhedrin sequences. Recircularization of this fragment would result in an non-viable genome. In contrast, recombination with a co-transfected transfer vector would result in the insertion of foreign sequences together with the reconstruction of the entire coding sequence of the essential 1629 ORF gene. Accordingly, only the recombinant viruses are, in theory, viable. In practice some parental virus is produced due to incomplete restriction with the enzyme. However, the frequency of recombinant viruses in the final progeny is generally very high when this method is used. Two such AcMNPV derivatives have been constructed, PAK5 and PAK6 (46). PAK5 contains the polyhedrin gene and recombination with the transfer vector leads to the production of virus with the occ⁻ phenotype whereas any contaminating parent virus would produce occ⁺ plaques. PAK6 is similar to PAK5 except that it encodes the *E. coli lacZ* gene instead of polyhedrin. Therefore, although both parent and the recombinant viruses would produce occ⁻ plaques, the recombinants will have lost the *lacZ* gene and therefore produce white plaques in the presence of β-galactosidase substrate. PAK6 contains an additional *Bsu*361 restriction site within the coding sequence of the *lacZ* gene. Therefore digestion of PAK6 viral DNA with *Bsu*361 will cut the genome at three sites within and around the polyhedrin driven *lacZ* gene, one of them cutting within the 1629 ORF. The advantage of having additional *Bsu*361 restriction sites is that it increases the probability of linearization of the parent genome DNA, thus allowing the transfection of almost 100% linear DNA rather than linear DNA contaminated with some uncut circular viral DNA (46). All transfer vectors currently available, other than those used for the yeast selection system (see Section 7.5), contain the 1629 ORF and can therefore be used in conjunction with PAK5 and PAK6. This latest modification greatly facilitates the isolation of recombinant viruses and, as described above, is the culmination of a number of improvements that have been made to the original procedure. Its use will be described in detail in later sections.

6.3 Yeast baculovirus system

A very different method for generating recombinant viruses rapidly and efficiently has been developed that takes advantage of the powerful and well-characterized genetics of the yeast, *Saccharomyces cerevisiae* (18). The yeast baculovirus system also overcomes many of the difficulties and time-consuming aspects of some of the previous methods for generating recombinants. The basis of the system is the creation of a modified baculovirus genome that can be maintained and propagated in yeast. A strain harbouring this genome is used as the recipient for the transfer vectors and the recombination necessary to produce the recombinant genomes takes place in this yeast background. Such homologous recombination is very efficient in yeast and a selection system has been incorporated to allow selection of yeast strains containing recombinant baculovirus genomes. Consequently, selection is rapid and efficient and in addition, the recombinants obtained are devoid of any background of parental, non-recombinant virus. Final purification of recombinants by one or more rounds of plaque isolation is therefore completely unnecessary. This method also lends itself well to the simultaneous isolation of multiple recombinants which should prove useful if, for example, a number of mutant forms of a protein need to be over-expressed for *in vitro* studies. Detailed protocols for the use of this method are also described in subsequent sections. Protocols for cloning foreign sequences into the transfer vectors can be obtained from the several cloning manuals currently available and are therefore not described here.

7. Using the baculovirus expression system

7.1 Introduction

Two methods for the isolation of recombinant baculovirus will be considered in detail in this chapter. The first method uses linearized parent virus which is co-transfected with a transfer vector containing the gene to be over-expressed. The medium from the transfected cells is used in a plaque assay to separate and isolate recombinant virus from any parent virus that may be present, by visual screening of recombinant plaques. The second method uses *S. cerevisiae* to select recombinants. DNA is then prepared from the selected yeast cells and used to transfect insect cells. Since no parent viral DNA is present, all the virus released in the medium following transfection will be recombinant and therefore there is no need to use plaque assays to select recombinants.

The cell lines commonly used for the propagation of AcMNPV and its recombinants are Sf9 and Sf21AE, both derived from IPLB-Sf-21, a *Spodoptera frugiperda* pupal ovarian cell line. Both cell lines grow well in monolayer and in suspension cultures and have a doubling time of 18–22 hours. The Sf9 cell line is available from American Type Culture Collection (accession number ATCC CRL 1711). Other insect cell lines such as TN-368 and BTI-TN-5B1–4 (also known as 'High 5') that also support the replication of AcMNPV are preferred by some laboratories. However, although they have been reported to produce higher yields of heterologous protein (Invitrogen Corporation), these cell lines do not grow well in suspension and as such have a limited use for large scale infections and production of proteins (3). Tissue culture media suitable for the growth of these cells, such as TC100, TNM-FH, Supplemented Grace's medium (containing yeastolate and lactalbumin hydrolysate), and IPL-41 are available from major suppliers. All the above media are supplemented with 10% fetal bovine serum (FCS). These media are buffered with sodium phosphate and have a pH of around 6.2. Incubation in an atmosphere of CO_2 is therefore not required. Several low serum and serum-free media for insect cell culture are also commercially available. Such media are preferred when the expressed protein is secreted. Antibiotics such as gentamycin (50 μg/ml) and amphotericin B (fungizone) (2.5 μg/ml) can be included if desired, to minimize the risk of bacterial or fungal contamination. Insect cells grow optimally at 27–28 °C but will grow at lower temperatures albeit at reduced growth rate.

Although the methods described here use Sf9 cells, the same methods equally apply to other Sf cell lines. For culture of Sf9 cells, we have used Supplemented Grace's insect cell medium (containing yeastolate and lactalbumin hydrolysate) (Life Technologies), but, other media described above are equally functional. 10% FCS was added to the medium to produce complete medium for cell growth.

7.2 Routine culture and storage of Sf9 cells

Sf9 cells grow well as monolayers and in suspension and can be easily transferred from one to the other.

7.2.1 Monolayer cultures

The cells are normally subcultured twice a week when they are 80–90% confluent (*Protocol 1*). Overgrown cultures will have a large proportion of floating cells (a small number of floating cells is a normal occurrence). Viewed under a phase-contrast microscope, the overgrown culture will be seen to contain a substantial proportion of unhealthy cells; healthy cells will appear round and shiny, whereas the dead cells appear dark and granular. If unsure about the state of the cells, cell viability can be assessed using a trypan blue exclusion test (see *Protocol 1*, step 4).

7.2.2 Suspension cultures

Sf9 cells grow easily in suspension and are convenient for large scale infections. Cultures of up to 1 litre will grow without the need for air-sparging. Suspension cultures are grown in the same medium as monolayer cultures and are subcultured whilst the majority of cells are still viable (*Protocol 2*).

Protocol 1. Subculturing monolayer cultures

Equipment and reagents

Most tissue culture laboratories will have all the equipment necessary for routine tissue culture procedures.

- Laminar air flow tissue culture cabinet
- 27 °C incubator
- Haemocytometer
- Complete insect cell culture medium (see Section 7.1) pre-warmed to 27 °C

- Tissue culture flasks or plates
- 0.2% (w/v) Trypan blue (Sigma) in tissue culture medium or phosphate-buffered saline (PBS)

Method

1. When the cells are ready for subculturing, examine them closely under the microscope to check for the health of the cells and for contamination. If the cells are in good condition, remove the medium from the tissue culture flask. Replace with an equal volume of fresh medium.

2. Dislodge the cells in the fresh medium by tapping the flask sharply against your hand, or by rapidly pipetting the medium over the cells, or gently scraping the cells with a sterile cell-scraper. Do not use trypsin or versine to detach the cells.

3. Pipette the cell suspension rapidly a few times to separate any clumps, taking care to minimize foaming.

4. Remove an aliquot and count the cells using a haemocytometer. Cell viability can be assessed at this stage by diluting with equal volume of 0.2% (w/v) trypan blue solution. Dead cells will take up the stain and appear dark blue while live cells will not.

5. In general, cells diluted 1:8 in growth medium will be ready for subculture three to four days later. Seeding densities for different size culture vessels are given below:

   - 25 cm^2 flask $1-2 \times 10^6$ cells in 5 ml of medium
   - 75 cm^2 flask 5×10^6 cells in 12 ml of medium
   - 150 cm^2 flask 1×10^7 cells in 25 ml of medium

6. Rock the flask gently to distribute the cells evenly. Incubate at 27 °C.

Protocol 2. Subculturing suspension cultures

Equipment and reagents

- Complete insect cell culture medium (see Section 7.1) pre-warmed to 27 °C
- 0.2% (w/v) Trypan blue (Sigma) in tissue culture medium or phosphate-buffered saline (PBS)
- Haemocytometer
- 250 ml spinner flask (e.g. Techne)
- Magnetic stirring platform (e.g. Techne)
- Water-bath suitable for the stirring platform (e.g. Techne) adjusted to 27 °C
- Sf9 cells

Method

1. Transfer pre-warmed (27 °C) medium to a spinner flask. Seed at a density of $2-3 \times 10^5$ Sf9 cells/ml.

2. Incubate at 27 °C with constant stirring at 60–80 r.p.m.

Protocol 2. *Continued*

3. The cells will be ready for subculturing when the density reaches $1-2 \times 10^6$ cells/ml and $> 97\%$ of the cells are viable (as judged by trypan blue staining; see *Protocol 1*, step 4). Cultures growing at higher densities are acceptable for passaging as long as the number of viable cells is $> 80\%$, but do not use these cells for infection. If the cells start to clump during growth, increase the speed of stirring.

7.2.3 Long-term storage of cells

Sf9 cells are most easily stored long-term after freezing in DMSO. A suitable procedure is described in *Protocol 3*.

Protocol 3. Long-term storage of Sf9 cells

Equipment and reagents

- Sf9 cells growing in log phase (see *Protocols 1* and *2*)
- Complete medium (see Section 7.1) containing 10% FCS and antibiotics (50 μg/ml gentamycin and 2.5 μg/ml amphotericin B)
- Freezing medium: 20% DMSO made up in complete medium, filter sterilized
- Cryovials (Life Technologies)

Method

1. Select healthy ($> 97\%$ viable) cells growing in log phase for long-term storage at $-80\,°C$ or in liquid nitrogen.

2. Pellet the cells at 1000 *g* for 10 min in a bench centrifuge. Discard the supernatant and resuspend the cells in complete medium to give a density of 4×10^6 cells/ml.

3. Dilute the cell suspension with equal volume of freezing medium.

4. Dispense a 1 ml aliquot into each cryovial and transfer to an insulated box. Freeze slowly at $-20\,°C$ for 2 h and then place at $-70\,°C$ overnight.

5. Transfer the vials to liquid nitrogen the next day.

6. Check the viability of cells by thawing a vial after two to three weeks. Place the vial in a 27 °C water-bath to thaw the cells rapidly. Transfer the contents into a 25 cm^2 flask containing 5 ml of pre-warmed complete medium.

7. Incubate at 27 °C for 1–2 h until the cells have attached.

8. Remove the medium and replace with 5 ml fresh medium. Incubate at 27 °C until the cells are ready for subculturing.

7.3 Infection of cells with virus

This should always be carried out using cells in the logarithmic phase of growth, as described in *Protocol 4*. The virus titre after infection and incubation should be determined by plaque assay (*Protocol 5*).

Protocol 4. Infection of insect cells with virus

Equipment and reagents

- Equipment as for *Protocol 1*
- Serum-free medium (see Section 7.1) containing antibiotics (50 μg/ml gentamycin and 2.5 μg/ml amphotericin B)
- Above medium supplemented with 10% FCS

Method

1. Always use healthy actively-growing cells for infection with virus. Count the cells and seed at the densities given below in medium without serum:
   - 35 mm plate 1.5×10^6 cells in 2 ml medium
   - 60 mm plate 2.5×10^6 cells in 4 ml medium
   - 25 cm^2 flask 2.5×10^6 cells in 5 ml medium
   - 75 cm^2 flask 8×10^6 cells in 12 ml medium

2. Allow the cells to attach for 30–60 min.

3. Remove the medium and add virus at a multiplicity of 10 p.f.u./cell. Rock the plate/flask gently to distribute the inoculum evenly. Incubate for 1 h, rocking the plates/flasks every 15 min gently to ensure coverage of cells with virus.

4. Remove the inoculum carefully without dislodging any cells.

5. Replace with an appropriate volume of complete medium and incubate the culture vessel for the required period of time.

Protocol 5. Routine plaque assay

Equipment and reagents

- 40°C water-bath
- Tenfold dilutions of virus (or supernatant from transfections) in tissue culture medium containing 10% FCS and antibiotics (50 μg/ml gentamycin and 2.5 μg/ml amphotericin B)
- 60 mm tissue culture dishes
- 50 ml 2 × complete medium (see Section 7.1) supplemented with 10% FCS, 50 μg/ml gentamycin, and 2.5 μg/ml amphotericin B
- 50 ml 3% (w/v) low melting point agarose (SeaPlaque) in distilled water in a 100 ml bottle, sterilized by autoclaving
- 0.2% (w/v) neutral red (Sigma), filter sterilized
- (*Optional*; only required if *lacZ* expression is used for screening) 20 mg/ml X-Gal (5-bromo-4-chloro-3-indolyl-β-D-galactopyranoside, Sigma) made up in dimethylformamide

Protocol 5. *Continued*

Method

1. The plaque assay requires 60 mm tissue culture dishes pre-seeded with 2.5×10^6 cells/dish. To ensure even distribution of the Sf9 cells, calculate the number of cells required and resuspend in the total volume of serum-free medium allowing 4 ml of medium per dish. Aliquot 4 ml of cell suspension per dish. Allow the cells to attach for 30–60 min at 27 °C.

2. Aspirate the medium from the tissue culture dishes. Replace with 1 ml of diluted virus. Rock the plates to distribute the virus evenly. Tenfold dilutions between 10^{-5} and 10^{-8} should be used for virus titration. Incubate at room temperature for 60 min to allow adsorbtion of virus.

3. While the virus is adsorbing to the cells, prepare the agarose overlay. To do this, melt the low melting point agarose in a microwave oven. Then transfer the molten agarose and the 2 × complete medium to a 40 °C water-bath. Keep these at 40 °C until required. Just before use, pour the medium into the bottle containing the agarose. Mix by swirling and replace in the water-bath.[a] It is important that the agarose has cooled to 40 °C before mixing with the medium.

4. Aspirate all the virus inoculum from the cells and overlay with 4 ml of agarose mixture (from step 3), taking care not to disturb the cell layer by letting the agarose run gently down the side of the dish. Rock the plates gently immediately after adding the overlay.

5. Allow the overlay to solidify at room temperature for 30–60 min.

6. Transfer the plates to a humidified incubator and incubate at 27 °C for four to seven days. If the incubator used is not humidified, place the plates in a large sandwich box containing some damp tissue placed on one side.

7. After four days, inspection of the plates will reveal small plaques. If titering wild-type virus, occ[+] plaques will be clearly visible. For recombinant virus it may be necessary to incubate the plates for a few more days. In this case, prepare more agarose overlay as before (step 3) but using half the volumes. Overlay each plate with 2 ml of agarose. Allow the overlay to solidify and re-incubate the plates for a further two to four days.[b]

8. The plates should be ready for counting plaques five to seven days p.i. Hold the plate inverted against a dark background and count the plaques. Multiply the figure by the dilution to give the titre of the virus stock.

[a] If *lacZ* expression is used for screening, add X-Gal to a final concentration of 120 µg/ml to the agarose overlay mixture.
[b] Addition of a second overlay four days post-infection helps to maintain the cell monolayer and in visualizing the plaques formed. Neutral red can also be used for visualizing plaques.

7.4 Production of recombinant viruses using linearized parental DNA

7.4.1 Transfer vectors

The past few years has seen the development of several transfer vectors for the expression of either one or multiple heterologous genes. AcMNVP-based transfer vectors which facilitate the insertion of foreign genes at appropriate sites within the polyhedrin or p10 genes have been developed. These may be used to produce either a fusion protein (containing some N-terminal poly-hedrin or p10 protein sequence) or mature, non-fusion, recombinant protein. The polyhedrin promoter is easily the most popular baculovirus promoter used to direct the expression of the foreign protein in the baculovirus ex-pression system. Some of the currently available vectors based on this pro-moter will be described here. The most up-to-date vectors contain the entire untranslated polyhedrin leader sequence and are engineered to allow insertion of foreign cDNA with or without a translation ATG start codon.

The transfer vector pAcYM1 (*Figure 2a*) contains the entire polyhedrin promoter and 5′ untranslated sequences up to the + 1 position (+ 1 being the translational start site) and has a single *Bam*HI restriction site for insertion of foreign coding sequences (47). The polyhedrin ATG has been destroyed and hence an ATG codon must be supplied with the insert. It is important when using these vectors that little if any foreign untranslated sequences are in-serted along with the coding sequence. Approximately 3 kb of flanking baculoviral sequences are present on either side of the insertion point to facilitate homologous recombination. In the case of pEVmXIV (48) (*Figure 2b*) which also requires an ATG start codon, the heterologous gene is ex-pressed under the control of a modified polyhedrin promoter which has been reported to increase the efficiency of polyhedrin transcription (approximately 50%) compared to the wild-type promoter. A polylinker is inserted at the + 1 position. The transfer vector pVL1393 (49) (*Figure 2c*) has the natural poly-hedrin translation codon mutated from ATG to ATT and foreign cDNA sequences are inserted at unique cloning sites located at position + 35 within the polyhedrin sequences. This modification was made because it was thought that sequences downstream of the polyhedrin ATG may contribute to optimal expression of the gene. However, recent studies have found that some translation does appear to occur from the ATT (50) and if an authentic non-fusion protein is required, care should be taken to ensure that the ATG of the inserted gene is out of frame with the ATT codon. Whether this inter-feres with the optimum expression of the gene of interest remains to be assessed.

In contrast to the above vectors, pAcC4 (*Figure 2d*) (described in ref. 7) supplies an ATG codon. Mutations at the − 2 and − 1 positions have resulted in an ATG translational start codon being embedded in a unique *Nco*I site. Thus heterologous genes lacking an ATG codon can be inserted at this *Nco*I

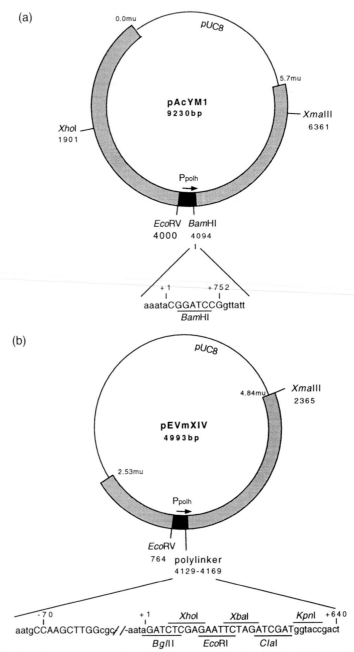

Figure 2. Transfer vectors used to obtain recombinant baculovirus by homologous re-combination in insect cells. See the text for detailed descriptions of individual vectors. In all cases, the dark filled area indicates the polyhedrin promoter (P_{polh}), the grey area the baculovirus sequences, and the thin line the plasmid backbone.

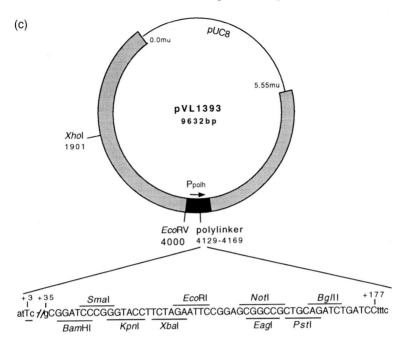

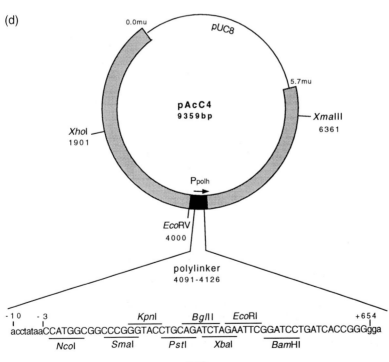

223

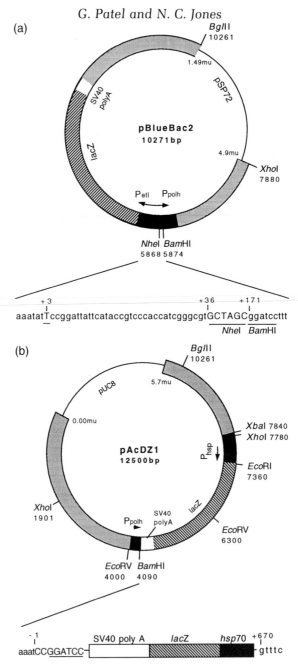

Figure 3. Transfer vectors utilizing the *E. coli lacZ* gene for screening purposes. See the text for detailed descriptions of the two vectors shown. In both cases, the dark filled area indicates the polyhedrin promoter sequences, the grey area the baculovirus sequences, and the thin line the plasmid backbone. P_{polh} is the polyhedrin promoter, P_{etl} is the promoter of the *etl* gene, and P_{hsp} is the *Drosophila hsp70* promoter.

site or at sites further downstream provided that the reading frame of the insert is in frame with the ATG.

Identification of recombinant occ⁻ plaques is a major hindrance for the unaccustomed researcher. Thus, to facilitate screening of polyhedra negative plaques, several vectors have been developed in which the *E. coli lacZ* gene is carried into the recombinant virus by the transfer vector in addition to the polyhedrin promoter driven foreign gene. Transfer vectors of this category include pBlueBac2 (Invitrogen) (*Figure 3a*) in which the *lacZ* gene is driven by a second viral promoter, *etl* (43), and pAcDZ1 (*Figure 3b*) in which the *lacZ* gene is under the control of a non-viral *Drosophila hsp70* promoter (44), which is active in insect cells. The recombinant virus arising through homologous recombination will also receive this *lacZ* gene and can therefore be identified by screening plaques for β-galactosidase production and for an occ⁻ phenotype. Like pVL1393, pBlueBac2 contains the natural polyhedrin translation codon mutated to ATT and a site for the insertion of foreign cDNA sequences located at +35 within the polyhedrin sequences. Care should therefore be taken to avoid the production of fusion protein initiated at the ATT codon.

7.4.2 Linearized parental genomic DNA

Recombinants are produced by the introduction into insect cells of linearized parental genomic DNA together with the appropriate transfer vector containing the inserted foreign sequence. The resulting mixed population of virus containing parent virus particles and recombinant virus are used in a plaque assay which distinguishes recombinant virus by different plaque characteristics. Any of the viruses that contain a *Bsu*361 site can be used as the parental genomic DNA in this method although PAK5 and PAK6 (see Section 6) give the best frequencies of recombinants in the final progeny. *Protocol 6* describes the preparation of linear genomic DNA.

Protocol 6. Purification of infectious baculovirus DNA

Equipment and reagents

- Virus for infection (e.g. PAK5, PAK6; see Section 6)
- Beckman ultracentrifuge, SW28 and SW40 rotors and tubes (or equivalent equipment)
- TE buffer: 10 mM Tris–HCl pH 8.0, 1 mM EDTA
- 10% and 50% (w/v) sucrose made up in TE buffer
- 20% (w/v) Sarkosyl made up in TE buffer
- Proteinase K (Sigma)
- Sf9 cells: 200–500 ml suspension culture at 5 × 10⁵ cells/ml
- Sterile centrifuge tubes (Becton Dickinson)
- Phenol/chloroform/isoamyl alcohol (25 : 24 : 1)
- CsCl
- 10 mg/ml ethidium bromide
- Butanol
- *Bsu*361 (New England Biolabs)
- 0.6% agarose gel in TBE buffer (89 mM Tris, 89 mM boric acid, 2 mM EDTA) plus horizontal gel electrophoresis equipment and electrophoresis buffer

Protocol 6. *Continued*

Method

1. Grow Sf9 cells to a density of 5×10^5 cells/ml in a suspension culture flask. Infect with virus (PAK5, PAK6, or any other virus) at a multiplicity of infection (m.o.i.) of 0.1–0.2 p.f.u./cell for four to five days until all the cells look infected when inspected under the microscope. Infected cells will appear rounded and significantly larger than the uninfected control cells.

2. Transfer the cell suspension to sterile 50 ml (or equivalent) tubes and spin down the cells at 2500 r.p.m. in a bench centrifuge.

3. Transfer the supernatant to ultracentrifuge tubes and pellet the virus particles at 80 000 g (24 000 r.p.m. in an SW28 rotor) for 1 h at 4 °C.

4. Decant the supernatant carefully and resuspend the faint translucent pellet of virus particles in a total of 2 ml of TE buffer.[a]

5. Prepare a 10–50% (w/v) one-step sucrose gradient in an SW40 tube. Prepare the gradients by layering 5.5 ml of 10% (w/v) sucrose over an equal volume of 50% (w/v) sucrose. Layer the virus over the sucrose and centrifuge at 100 000 g for 1 h at 4 °C.

6. After centrifugation, the virus band is visible at the interface of the two sucrose solutions. Harvest the virus band with a Pasteur pipette.

7. Dilute the virus with TE buffer, and pellet it in an ultracentrifuge in an SW40 rotor at 100 000 g for 1 h.

8. Allow the pellet to resuspend in 400 µl of TE buffer overnight.

9. Add 100 µl of 20% Sarkosyl in TE buffer. Incubate at 60 °C for 30 min.

10. Add proteinase K to a final concentration of 500 µg/ml. Incubate at 37 °C for 2–16 h.

11. Extract DNA twice with an equal volume of phenol/chloroform/isoamyl alcohol (25 : 24 : 1). Do not vortex. Invert the tube quickly to mix the two phases.

12. Ethanol precipitate the DNA.

13. Resuspend the DNA in 5 ml of TE buffer. Add 5 g of CsCl, mix to dissolve, and add 25 µl of 10 mg/ml ethidium bromide. Spin in an ultracentrifuge at 200 000 g overnight at 20 °C.

14. Harvest the DNA band and extract the ethidium bromide with butanol.

15. Ethanol precipitate the DNA or dialyse the DNA against TE buffer.

16. Measure the OD_{260} of the DNA solution to determine the concentration of DNA.

17. Digest 2 µg of DNA in a 100 µl volume with *Bsu*361. Remove 5 µl and check on a 0.6% agarose gel to verify that the DNA is cut.

18. Heat inactivate the residual enzyme at 65 °C for 10 min. Use 10 µl for co-transfections of Sf9 cells with 2 µg of transfer vector containing the gene to be expressed.

a The pellets can be left in TE buffer overnight at 4 °C to aid resuspension.

7.4.3 Co-transfection of linear viral DNA and transfer vector

Three techniques have been described for the introduction of DNA into insect cells; cationic lipid-mediated transfection (lipofection), transfection with a calcium phosphate–DNA co-precipitate, and electroporation. In all cases the quality of viral and plasmid DNA is critical for successful transfections. DNA that is freshly prepared and purified on caesium chloride density gradients should be efficient in transfection protocols.

Transfection of DNA into tissue culture cells by calcium phosphate precipitation is by far the most commonly used procedure and is described in *Protocol 7*. Recently, however, because of the ease of introducing DNA into cells and the efficiency and reproducibility of results obtained, lipofection is also widely used for insect cell transfections. This method is described later in *Protocol 13*.

All transfections should be performed in duplicate. In addition, if using parent virus DNA which is devoid of a functional polyhedrin gene (e.g. PAK6), it is recommended to use wild-type DNA as a positive control for transfections. Polyhedra produced in the positive control should indicate the success of the transfection procedure. A negative control can also be included which lacks parent virus DNA.

Protocol 7. Calcium phosphate transfection

Equipment and reagents

- 25 cm^2 tissue culture flasks each pre-seeded with 1 × 10^6 Sf9 cells
- Linear parental viral DNA prepared as in *Protocol 6*
- Plasmid transfer vector encoding the foreign DNA sequence of interest
- Transfection buffer: 25 mM Hepes pH 7.1, 140 mM NaCl, 125 mM CaCl$_2$—filter sterilized
- Grace's medium (Life Technologies) containing 10% fetal bovine serum
- Complete growth medium (see Section 7.1)

Method

1. Aspirate the medium from each tissue culture flask and replace with 0.75 ml Grace's medium. Leave the flask at room temperature.

2. To a 1.5 ml microcentrifuge tube, add 200 ng of linearized viral DNA, 2 µg of transfer vector, and 0.75 ml transfection buffer. Mix and add this dropwise with swirling to the Grace's medium in the flasks. Calcium phosphate precipitates due to CaCl$_2$ in the transfection buffer and phosphate in the medium.

Protocol 7. *Continued*

3. Incubate the flasks for 4 h at 27 °C.

4. Aspirate the medium from the flasks and wash the cell layer gently with complete medium. Add 5 ml of fresh medium and incubate the flasks for four to five days or until polyhedra are visible if the parent virus contained an intact polyhedrin gene.

5. Collect the medium and centrifuge at 1000 *g* for 5 min to remove cellular debris. Store the supernatant, containing the extracellular virus, at 4 °C.

7.4.4 Visual screening for recombinant virus

Recombinant viruses need to be isolated from contaminating parental virus in the pool of progeny obtained from the co-transfection. This is generally done by performing a plaque assay whereby plaques produced by the recombinant are phenotypically different from those produced by the parent. Visual screening of the plaques can rely on the presence or absence of one or both of the marker genes, polyhedrin and *lacZ*. The presence of a polyhedrin gene will result in the formation of occlusion bodies in the infected Sf9 cells while a functional *lacZ* gene will encode β-galactosidase which will produce a blue coloration in the presence of X-Gal. If using wild-type AcMNPV, or derivatives such as PAK5 (which produces occ$^+$ plaques) in conjunction with transfer vectors such as pAcYM1, pEVmXIV, pVL1393, or pAcC4 (*Figure 2*), the recombinant virus will produce occ$^-$ plaques because of the replacement of the polyhedrin gene with the foreign gene. However, if the same parent DNA is used with either pAcDZ1 or pBlueBac2 (*Figure 3*), the recombinants will not only be occ$^-$, but will also contain *lacZ* which allows the formation of blue plaques. In both of these cases, the recombinants will be produced in the background of wild-type, white, occ$^+$ plaques. A similar strategy is used for identifying recombinants when using parent virus which already has its polyhedrin gene replaced with the bacterial *lacZ* gene. For example, when using PAK6 virus (which produces blue, occ$^-$ plaques) in combination with vectors such as pAcYM1, pEVmXIV, pVL1393, or pAcC4, the viral plaques, whether recombinants or non-recombinants, will form occ$^-$ plaques. However, whereas the parent virus will form blue plaques the recombinants will produce white plaques because of the replacement of *lacZ* gene with the foreign gene. The proportion of recombinant plaques varies depending upon the parental virus that is being used. Best results are obtained with PAK6.

The procedure for visual screening of recombinant virus is described in *Protocol 8*. Since it relies on the plaque phenotype, the conditions required for the generation of good plaque assays is critical. Ideally, dilutions of virus should be assayed in duplicate. Cells must be healthy at the time of infection. The density of the cells on a plate is important especially when screening for recombinant viruses. Too many cells will result in formation of a complete monolayer with very small plaques which will be difficult to see. If the cells are too sparse, plaques may not form at all. If the experimenter is unfamiliar with the occ⁻ and occ⁺ plaque morphology, viruses eliciting these phenotypes should be obtained first in order to become familiar with visual identification under a microscope. Positive and negative controls should always be included for each plaque assay.

Protocol 8. Isolation of recombinant plaques

Equipment and reagents

- Good quality microscope (e.g. stereo dissecting microscope). For screening occ⁺/occ⁻ plaques, this should have a piece of black velvet on the platform as background and a strong light source at a very acute angle to the platform. In screening for the expression of the *lacZ* gene, use a white background instead.
- Tenfold dilutions of supernatant from transfections in tissue culture medium (see *Protocol 7*, step 5)

- 60 mm tissue culture dishes pre-seeded with 2×10^6 Sf9 cells/dish
- 50 ml $2 \times$ complete growth medium (see Section 7.1) supplemented with 10% serum and antibiotics
- 50 ml 3% low melting point agarose (Sea-Plaque) in distilled water in a 100 ml bottle sterilized by autoclaving
- For visualization of *lacZ⁺* plaques only: 20 mg/ml X-Gal (Sigma) made up in dimethylformamide

Method

1. Follow *Protocol 5* up to step 6.

2. After four days incubation at 27 °C, inspection of the plates will reveal small occ⁺ plaques. Prepare an agarose overlay as described in *Protocol 5*, step 3, using half the volumes. Overlay each plate with 2 ml of agarose. Allow the overlay to solidify and re-incubate the plates for a further two to four days.[a]

3. To visualize *lacZ⁺* plaques, overlay the plates with 1 ml of complete medium containing 20 µl of the X-Gal solution. Incubate the plates at 27 °C for 6–16 h. Remove the medium after the incubation period.

4. The plates should be ready for visual screening five to seven days p.i. Carry this out as follows, depending on whether occ⁻ plaques or *lacZ⁺* plaques are desired.

    (a) Occ⁻ plaques. Inspect the plates using the microscope with black velvet on the platform. Against the black background the occ⁺ plaques will be clearly visible as clusters of very refractile, shiny cells against a dull background of uninfected cells. Occ⁻ plaques lack the occlusion bodies and are therefore not refractile and will require a greater scutiny.[b] Infected cells become rounded and, compared to uninfected background monolayer, occ⁻ plaques will appear whitish with a centre that is devoid of cells. However, care should be taken not to mistake discontinuity in the cell monolayer as a recombinant occ⁻ plaque.

Protocol 8. *Continued*

    (b) *lacZ*[+] plaques. Inspect the plates using the microscope with a white background to help in visualizing the blue coloration. If the colour is too faint to distinguish, the entire plate could be frozen and thawed two or three times which will lyse the cells, exposing the β-galactosidase to X-Gal.

5. Mark the prospective candidates for picking and examine them under a phase-contrast microscope. If picking for recombinants lacking in the production of polyhedrin protein, ensure that none of the occ⁻ plaques marked contain occlusions. Eliminate any false positives.

6. Pick the confirmed recombinant plaques using a Pasteur pipette or a micropipette. Place the tip directly on to the plaque and remove a small plug of agarose overlay into 1 ml of medium.

7. Vortex the agarose plug contained in the medium to release the virus. These isolates should be re-plaqued and purified several times using the same method described above.

---

[a] Very small occ⁺ plaques are visible three to four days post-infection. However, in order to see occ⁻ plaques clearly we recommend incubating the plates for a further two to four days (p.i.).
[b] 0.2% (w/v) neutral red can be used for visualizing plaques. However, since neutral red stain becomes granular on exposure to light, occ⁻ plaques which have taken up the stain may be mistakenly scored as occ⁺ plaques. It is therefore not recommended to use neutral red for initial isolation of occ⁻ plaques.

7.5 Isolation of recombinant baculovirus in yeast

7.5.1 Strategy

The strategy employed for the rapid and efficient isolation of recombinant baculoviruses using a yeast system is outlined in *Figure 4*. It is based on the ability to select directly viral genomes that have undergone homologous recombination (18). YGP2 yeast cells contain a baculovirus derivative that is stably maintained in yeast. Immediately downstream of the polyhedrin promoter a cassette of yeast DNA elements is inserted consisting of a yeast ARS (autonomous replication sequence), a CEN (centromere) sequence that ensures the maintenance of the baculovirus DNA during mitosis, and two selectable marker genes, *URA3* and *SUP4-o*. Viral DNA containing these elements is stably maintained in yeast but remains infectious for insect cells. Yeast harbouring such DNA provides an ideal environment for homologous recombination to generate appropriate recombinant baculovirus DNA. The *SUP4-o* gene is located downstream of the polyhedrin promoter. Homologous recombination between the genome and transfer vector results in the loss of *SUP4-o* gene and its replacement with the polyhedrin promoter driven foreign gene. Therefore recombinants can be selected by a *SUP4-o* counter-

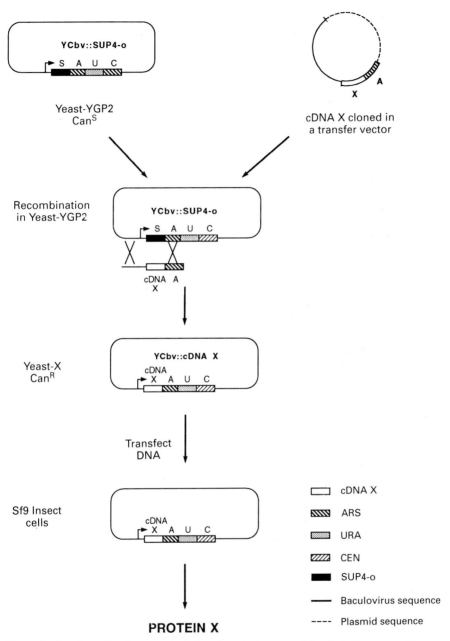

PROTEIN X

Figure 4. A schematic diagram depicting the strategy used for isolating recombinant baculovirus using *S. cerevisiae*. See the text for details. YCbv::SUP4-o represents baculovirus DNA contained within the yeast YGP2. YCbv::cDNA X represents recombinant baculovirus DNA, which contains X cDNA sequences instead of the *SUP4-o* gene, after homologous recombination in yeast.

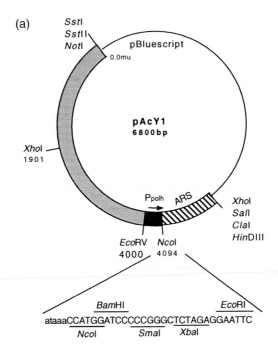

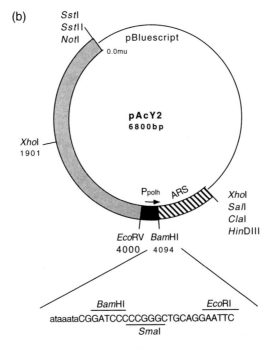

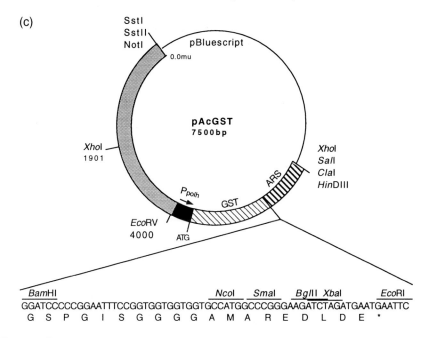

Figure 5. Three transfer vectors used to obtain recombinant baculovirus by homologous recombination in yeast cells. See the text for details of individual vectors. In all cases, the dark filled area represents the polyhedrin promoter (P$_{polh}$), the grey area the baculovirus sequences upstream of the polyhedrin promoter, and the dark striped area the yeast ARS sequences. The light striped area in (c) denotes the glutathione S-transferase sequences (GST).

selection. DNA from the selected yeast colonies can then be isolated and used to transfect insect cells to allow the production of recombinant virus.

Three vectors have been designed for the use of the yeast/baculovirus method for the selection of recombinant virus. pAcY1 (*Figure 5a*) is based on the pAcC4 transfer vector (*Figure 2d*) and contains the polyhedrin promoter and leader sequence followed by an *Nco*I site which provides a translation initiation codon (18). pAcY2 (*Figure 5b*) (18) is based on pAcYM1 (*Figure 2a*) and therefore will need an ATG inserted with the heterologous cDNA. In contrast to previously described vectors where the insert is flanked on both sides by baculovirus DNA, these two vectors, pAcY1 and pAcY2, result in the insert being flanked 5′ with viral sequences and 3′ by yeast ARS (autonomous replicating sequences). pAcY2 has been further modified to yield pAcYGST (*Figure 5c*), encoding the *Schistosoma japonicum Sj26* antigen, glutathione S-transferase (GST). The GST sequence is located immediately downstream of the polyhedrin promoter and the entire untranslated leader sequences. It is followed by a thrombin cleavage site and several unique cloning sites allowing the incorporation of the cDNA of interest to be fused in-frame with the coding sequences of GST. The protein is expressed as a GST fusion which can be isolated in a single step using glutathione–Sepharose beads. This may be used directly for some experiments as an immobilized matrix or alternatively eluted from the beads by the addition of excess glutathione. Subsequently, the expressed protein can be separated from the GST moiety by cleavage with thrombin. The advantage of using this modified vector system is the ease by which the over-expressed protein can be purified from the infected cells.

The strain YGP2 (18) is a derivative of the yeast strain y657 (MAT a his3-11,15 trp1-1, ade2-1, leu2-3,112 ura3-52, can1-100, his4::HIS3) (51), where the ochre suppressable mutations on ade2-1 and can1-100 alleles are suppressed by the *SUP4-o* gene. Hence, these yeast cells are able to grow in the absence of exogenous adenine. In addition, YGP2 cells are able to synthesize an arginine permease encoded by the can1 gene, since the can1-100 mutation is also suppressed. This permease not only facilitates the uptake of arginine but can also facilitate the uptake of certain arginine analogues such as the toxic analogue, canavanine. Thus, YGP2 cells are unable to grow in the presence of canavanine. Growth in the absence of adenine and sensitivity to canavanine are therefore the two phenotypes observed for the presence of the suppressor gene, *SUP4-o*. Loss of the *SUP4-o* gene (and hence replacement with the foreign gene of interest) will elicit an entirely different phenotype since the two mutations, ade2-1 and can1-100, are no longer suppressed. The cells are unable to produce a functional arginine permease enzyme and are therefore resistant to the presence of canavanine in the growth medium. Furthermore, the addition of exogenous adenine in the growth media is required; in this case, since there is a block in the pathway of adenine biosynthesis, an intermediate pink product is produced rendering the yeast colonies pink. The phenotypes elicited by the presence or the loss of *SUP4-o* gene are depicted in *Table 1*.

Table 1. Phenotypes of YGP2 and its derivatives[a]

Yeast strain	Replicating genome	Partial genome	Phenotype
YGP2	YCbv:SUP4-o	can1-100, ade2-1, SUP4-o	Ade+ Can^S
YGPX	YCbv:X	can1-100, ade2-1	Ade- Can^R

[a] This table shows phenotypes of YGP2 and its derivatives that contain or lack the *SUP4-o* gene. YGP2 contains episomally maintained baculovirus DNA with the yeast *SUP4-o*, ARS, URA3, and CEN genetic elements inserted downstream of the polyhedrin promoter (*Figure 4*). YGPX is a derivative of YGP2 where the replicating genome contains the cDNA sequences of the gene X, to be over-expressed, in place of the *SUP4-o* gene.

7.5.2 Transformation of yeast cells

Transfer vector containing the foreign gene is introduced into yeast sphaero-plasts in conjunction with a plasmid containing TRP gene (e.g. RS314, ref. 18). The preparation of yeast sphaeroplasts and the co-transformation procedure is described in *Protocol 9*. Transformants are initially selected on the basis of their TRP$^+$ phenotype. Only the larger, faster growing colonies are selected for further analysis.

Protocol 9. Introduction of DNA into yeast cells

Equipment

- 30 °C incubator
- A small orbital shaker set at 30 °C
- Sterile 15 ml tubes (Becton Dickinson)
- Plastic Petri dishes
- Sterile toothpicks
- Conical flasks

Media for yeast culture

All of the following solutions should be autoclaved or filter sterilized and aseptic techniques should be employed to avoid bacterial contamination. Add 0.1 vol. of 20% (w/v) glucose and relevant supplements before use.[a]

- SD minimal medium: 6.5 g/litre yeast nitrogen base (without amino acids) (Difco)
- −UT (−ura, −trp) medium: SD minimal medium supplemented with 20 µg/ml histidine (Sigma), 20 µg/ml adenine (Sigma), 60 µg/ml leucine (Sigma).
- −U (−URA) medium: UT (−ura, −trp) medium supplemented with 20 µg/ml tryptophan (Sigma), and 11 mg/ml casamino acids (Difco)

- −U/CAN: −U medium without casamino acids supplemented with 60 µg/ml canavanine sulfate (Sigma)
- Agar plates: 2% Bacto agar added to −U, −UT, −U/CAN media to produce the relevant agar plates
- Top agar: 1 M sorbitol added to SD agar before autoclaving
- 20% (w/v) glucose

Reagents for yeast transformation

- 1 M sorbitol
- SCE solution: 1 M sorbitol, 0.1 M sodium citrate pH 5.8, 10 mM EDTA
- Lyticase (Sigma): dissolve lyticase in 1 ml SCE in a microcentrifuge tube, spin down the undissolved lyticase for 5 sec, and use the supernatant
- β-mercaptoethanol
- 5% (w/v) SDS
- STC solution: 1 M sorbitol, 10 mM Tris–HCl pH 7.5, 10 mM CaCl$_2$
- PEG solution: 10 mM Tris–HCl, pH 7.5, 10 mM CaCl$_2$, 20% polyethylene glycol 8000, filter sterilized
- SOS buffer: 1 M sorbitol, 6.5 mM CaCl$_2$, 0.25% yeast extract, 0.5% Bacto peptone, filter sterilized
- SOS buffer containing 20 µg/ml histidine, 20 µg/ml adenine, and 60 µg/ml leucine[a]

- −UT agar plates (see above)
- Top agar containing 2% glucose, 20 µg/ml histidine, 60 µg/ml leucine, and 20 µg/ml adenine
- 10 µg of transfer vector containing the foreign gene. This vector must be digested at two convenient restriction sites, one upstream of the polyhedrin promoter and the other downstream of the yeast ARS. It is unnecessary to isolate the released fragment. The cut plasmid should be ethanol precipitated and resuspended in 10 µl volume. Use 5 µg plasmid DNA for each transformation.
- Co-transforming plasmid containing the TRP gene (e.g. RS314, ref. 18). Use 0.5 µg of co-transforming plasmid DNA for each transformation.

Protocol 9. *Continued*

Method

1. Pick a colony of YGP2 and streak on to a −U agar plate. Incubate for one to two days until the colonies have grown to approximately 1 mm in diameter.

2. Using a sterile toothpick, pick a colony and plate on a −U/CAN agar plate. This is conveniently done by touching the plate with the tooth-pick to form a small line of approximately 1 cm. Using the same toothpick, which should still contain some yeast cells, plate on to a −U plate. Repeat this process using four more clones and the same −U/CAN and −U agar plate. Include a positive control (which is able to grow on both −U and −U/CAN plate) if available. Incubate at 30 °C for two days.

3. Check for growth on both the −U and −U/CAN plates. YGP2 is sensitive to canavanine and will fail to grow on the −U/CAN plate. Any clones that grow on both the plates should be discarded. Pick a clone from the −U plate which failed to grow on the −U/CAN plate.

4. Inoculate 50 ml of −U medium with approximately 1 mm^2 of the fresh canavanine-sensitive YGP2 patch from above or a single colony from a freshly streaked −U agar plate. Grow overnight at 30 °C in a shaking incubator to an OD_{600} of 0.5–0.8.[b]

5. Pellet the cells in a bench-top centrifuge for 5 min at 2500 r.p.m. Wash the cell pellet by centrifugation first with water and then with 20 ml 1 M sorbitol.

6. Pellet the cells again by centrifugation. Resuspend the pellet in 20 ml SCE solution and add 40 μl β-mercaptoethanol and 1000 U of lyticase. Incubate at 30 °C for 20–30 min with occasional shaking. Monitor the formation of sphaeroplasts during incubation by removing 5 μl of lyticase-treated cells, mixing with 5 μl of 5% SDS on a microscope slide and observing the formation of 'ghosts'. Compare with sphaero-plasts that have not been treated with SDS.

7. When 90% of the cells have formed sphaeroplasts (about 20 min), harvest the cells by pelleting at 1000 r.p.m. in a bench-top centrifuge for 5 min. Resuspend the sphaeroplasts gently in 20 ml of STC solution and spin down again. Resuspend the pellet in 2 ml of STC solution; 0.1 ml sphaeroplast suspension is needed for each transformation.

8. Mix 5 μg of restricted plasmid containing the foreign gene and 0.5 μg of co-transforming plasmid containing the TRP gene (e.g. pRS314, ref. 18) in a sterile 15 ml tube. Add 0.1 ml of sphaeroplast suspension (from step 7) to the tube and mix gently. Incubate the suspension at room temperature for 10 min.[c]

9. After 10 min incubation, gently suspend the sphaeroplast–DNA mix-ture in 1 ml of PEG solution and incubate at room temperature for a further 10 min.

10. Pellet the sphaeroplasts in a bench-top centrifuge at 1000 r.p.m. for 5 min. Remove the supernatant and resuspend the pellet in 150 µl of SOS buffer (containing histidine, adenine, and leucine). Incubate at 30 °C without shaking for 20–40 min.

11. Add 8 ml of molten top agar, held at 45 °C, to each tube. After quickly mixing, pour the contents on to a −UT agar plate. Allow the agar to solidify and incubate the plates inverted in a 30 °C incubator for approximately three days.

12. Initially two sizes of TRP$^+$ colonies will appear on the incubated plate. Pick the larger colonies. Colony purify these by re-plating.

---

[a] All supplements (histidine, adenine, leucine, tryptophan, and canavanine) can be made up as 100 × stock solutions, filter sterilized, and added to media prior to use.
[b] In order to get the correct density of the yeast culture, a one in five dilution of the above culture could also be set-up.
[c] For a positive control, use 0.5 µg of co-transforming plasmid, pRS314, on its own; add 5 µg of sheared salmon sperm DNA as carrier DNA. As a negative control, use 5 µg of restricted plasmid without the TRP gene.

7.5.3 Selection of canavanine-resistant colonies

Large fast growing colonies initially selected on the basis of their TRP$^+$ phenotype are next screened for their ability to grow on canavanine plates (*Protocol 10*). Only those colonies in which the *SUP4-o* gene has been lost (because of replacement by the foreign gene) will be able to grow on the canavanine agar.

Protocol 10. Identification of canavanine-resistant colonies

Reagents

- −U agar plate (see *Protocol 9*)
- −U/CAN agar plate (see *Protocol 9*)

Method

1. Choose the larger, faster growing TRP$^+$ colonies for further analysis (since, generally, SUP$^-$ cells grow faster than SUP$^+$ cells). Divide a −U plate into 8–16 segments. Streak individual TRP$^+$ colonies on to the segments in order to isolate single colonies. Incubate at 30 °C.

2. Pick one or two isolated colonies from each segment of the −U plate and 'replica' onto a −U/CAN agar plate. Incubate at 30 °C for one to two days.

Protocol 10. *Continued*

3. Observe the plates. Only the canavanine-resistant colonies which have lost the *SUP4-o* gene (which is replaced by the foreign gene of interest) will grow on canavanine plates.[a] These colonies will also become pink if adenine is limiting in the medium.

---

[a] Canavanine-resistant colonies can be further checked for the presence of the foreign sequences by preparing DNA and analysing either by PCR or Southern blot hybridization. However, this should be unnecessary since the vast majority of canavanine-resistant colonies should contain the foreign gene.

7.5.4 Purification of yeast DNA from canavanine-resistant cells

The next step is to purify total yeast DNA from canavanine-resistant cells for the transfection of Sf9 insect cells. The isolation of total yeast DNA is carried out by the preparation and subsequent lysis of sphaeroplasts, ethanol precipitation of the DNA, and the treatment with RNase and Proteinase K. The procedure is described in *Protocol 11*.

Protocol 11. Preparation of total yeast DNA

Equipment and reagents

- Sorvall centrifuge and HB-4 rotor (or equivalent)
- Corex tubes (or equivalent)
- 1 M sorbitol
- Lyticase (Sigma) (see *Protocol 9*)
- SCE solution: 1 M sorbitol, 0.1 M sodium citrate pH 5.8, 10 mM EDTA
- β-mercaptoethanol
- Lysis buffer: 4.5 M guanidinium chloride, 0.1 M EDTA pH 8.0, 0.15 M NaCl, 0.05% Sarkosyl
- 10 × TE buffer: 100 mM Tris–HCl pH 7.5, 10 mM EDTA
- TE buffer: 10 mM Tris–HCl pH 7.5, 1 mM EDTA
- 10 mg/ml DNase-free RNase (Sigma)
- 10 mg/ml proteinase K (Sigma)
- −U medium (see *Protocol 9*)
- Phenol/chloroform/isoamyl alcohol (25 : 24 : 1)
- 3 M sodium acetate pH 5.2

Method

1. Inoculate 5 ml of −U medium with a single canavanine-resistant colony. Incubate in a shaker at 30 °C for 4–8 h. Use all 5 ml to inoculate a larger volume (250 ml) of −U medium. Grow overnight at 30 °C to an OD_{600} of 1–1.5.

2. Use this culture to prepare the sphaeroplasts with lyticase as described in *Protocol 9*.[a]

3. Spin sphaeroplasts at 2000 r.p.m. for 5 min in a bench-top centrifuge and remove the supernatant. Resuspend the sphaeroplasts gently in

1 ml of TE buffer. Add 15 ml of lysis buffer slowly, mixing gently by stirring with the pipette.[b] Incubate at 65 °C for 10 min with occasional swirling.

4. Cool to room temperature and add 16 ml cold ethanol. Mix and spin at 10 000 r.p.m. in a Sorvall HB-4 rotor for 10 min.

5. Drain the pellet well. Loosen the pellet by gently stirring with a pipette[c] taking care not to shear the DNA. Resuspend the loosened pellet in 10 ml of 10 × TE buffer.

6. Add 50 μl of 10 mg/ml DNase-free RNase. Incubate at 37 °C for 30–60 min.

7. Add 150 μl of 10 mg/ml proteinase K. Incubate at 65 °C for 30 min.

8. Extract twice with 10 ml of phenol/chloroform/isoamyl alcohol (25 : 24 : 1). Do not vortex. Invert the tube quickly several times to mix the two phases. Transfer the aqueous phase to a Corex tube. Add 0.1 vol. of 3 M sodium acetate pH 5.2, and 2 vol. of cold ethanol. Cool at −70 °C for 10 min.

9. Spin at 10 000 r.p.m. for 10 min in a Sorvall HB-4 rotor. Rinse the pellet with 70% ethanol. Discard the supernatant and lightly dry the pellet under reduced pressure. Re-dissolve the pellet in 1 ml of TE buffer.[d]

10. Use 1–5 μl of DNA suspension for PCR analysis to ascertain the presence of the inserted gene in the baculovirus genome.

[a] The complete formation of sphaeroplasts may take longer than in the transformation method (Protocol 9) especially if the culture of the canavanine-resistant cells has reached stationary phase.
[b] It is important not to add all 15 ml lysis buffer at once. Add a little at a time stirring with a pipette.
[c] Special handling is required to resuspend this pellet. Do not add all 10 ml of 10 × TE buffer at once. Add a little at a time stirring with a pipette.
[d] Resuspension of DNA in TE buffer may take more than 1 h. A 10 min incubation at 42 °C may help.

DNA prepared as in *Protocol 11* can be used directly to transfect insect cells. In this case, prepare the DNA from 50 ml of culture, halve the volumes of the buffers given in *Protocol 11*, and resuspend the final pellet in 200 μl of TE buffer. Then, use several different amounts of the DNA for transfections; e.g. 1 μl, 2 μl, 5 μl, 10 μl, and 12 μl volumes in a 24 μl DNA–lipofectin mixture (see *Protocol 13*).

However, we have noted some inconsistencies in the results obtained when using DNA isolated directly as described in *Protocol 11* due to the presence of some inhibitory substance in the total yeast DNA preparation. For this reason we recommend partially purifying the DNA on a 5–20% sucrose gradient (*Protocol 12*).

Protocol 12. Purifying DNA on a sucrose gradient

Equipment and reagents

- Beckman ultracentrifuge and SW40 rotor (or their equivalent)
- Fraction collector
- NTE: 200 mM NaCl, 10 mM Tris–HCl pH 7.5, 2 mM EDTA
- 5% (w/v) sucrose in NTE
- 20% (w/v) sucrose in NTE
- TE buffer (see *Protocol 11*)

Method

1. Prepare a 12 ml 5–20% (w/v) continuous sucrose gradient in a Beckman SW40 tube. Layer the prepared DNA (from *Protocol 11*) on top of the gradient.

2. Centrifuge at 220 000 g (35 000 r.p.m.) for 3 h at 20 °C in an SW40 rotor.

3. Using a fraction collector, pierce the bottom of the tube and collect 0.5 ml fractions in 1.5 ml microcentrifuge tubes.

4. Add 1 ml cold ethanol to each tube, chill for 10 min at −70 °C, and pellet the DNA in a microcentrifuge.

5. Wash the pellet by centrifugation with 70% ethanol. Dry the pellet lightly and resuspend it in 50 μl of TE buffer.

Check for the presence of the foreign gene in the purified DNA by PCR analysis using PCR oligo No. 1, 2, or 3 (*Figure 6*) and another corresponding to the inserted gene. Under the sucrose gradient conditions described in *Protocol 12*, the bulk of the baculovirus DNA migrates half-way down the tube. Therefore, 1 μl of alternate fractions corresponding to the middle of the tube can be used for PCR analysis.

7.5.5 Transformation of Sf9 cells with yeast DNA

The DNA prepared from yeast and partially purified on a sucrose gradient can now be used to infect a culture of Sf9 cells. This DNA is free of any contaminating parental viral DNA and thus a plaque assay to further isolate recombinants is not required. The DNA can be introduced into the Sf9 cells by calcium phosphate transfection (*Protocol 7*). However, a lipofection procedure can also be used (*Protocol 13*) and works well with such DNA.

Protocol 13. Transfections of Sf9 cells with yeast DNA

Equipment and reagents

- Serum-free medium (see Section 7.1)
- Complete medium (see Section 7.1)
- Sterile polystyrene tubes (12 × 75 mm, Becton Dickinson)
- Lipofectin (Life Technologies)
- 35 mm dishes pre-seeded with 1×10^6 Sf9 cells

Method

1. Prepare a diluted lipofectin solution by mixing two parts of lipofectin with one part of distilled water in a polystyrene tube. Each 35 mm dish of Sf9 cells will require 12 μl diluted lipofectin (for a 24 μl final transfecting mixture).

2. Set-up several transfections in the polystyrene tubes using three or four different amounts of DNA to optimize the procedure. For example, use 1 μl, 3 μl, or 10 μl of DNA, each time making the final volume to 12 μl with sterile distilled water.

3. Add 12 μl of diluted lipofectin reagent to the 12 μl DNA solution and mix gently. Leave at room temperature for 15 min.

4. Remove the medium from the tissue culture dishes and replace with 1 ml of culture medium lacking serum.

5. Add the DNA–lipofectin mixture. Swirl to mix and incubate at 28 °C for 4–24 h.

6. Add 1 ml of complete medium containing 10% serum and continue the incubation for three to five days or until the cells look visibly infected (see *Protocol 7*). All the virus released in the medium is recombinant baculovirus expressing the foreign protein of interest and can be amplified by re-infecting fresh cell cultures.

7. Use 1 ml of the supernatant to infect a 25 mm^2 tissue culture flask pre-seeded with 2.5×10^6 Sf9 cells as described in *Protocol 4*.

Once a recombinant virus has been isolated, the expression of the foreign gene following infection of fresh Sf9 cells can be addressed. Although some idea of the efficiency of expression can be obtained using untitred stocks, it is best to obtain a stock of recombinant virus that has been titred accurately. Insect cells can be infected at known m.o.i. and so an accurate picture of the accumulation of foreign protein can be obtained. The easiest way to do this is to analyse extracts of infected cells by SDS–PAGE. Good expression levels should allow the foreign proteins to be seen following Coomassie blue staining of the gel but lower levels of expression may require radiolabelling of infected cells followed by autoradiography and fluorography of the gel. To label proteins expressed in infected cells efficiently, it is best to use medium deficient in methionine or leucine (which is commercially available from Life Technologies). If an antibody to the foreign protein is available, expression can be assessed by Western blotting or immunoprecipitation from radiolabelled extracts. Detailed protocols for these methods can be found in several laboratory manuals (3–5).

In some cases, the foreign protein is degraded at late times post-infection and it may therefore be necessary to harvest cells at earlier times after infection. The disadvantage to this is that at earlier times, the levels of

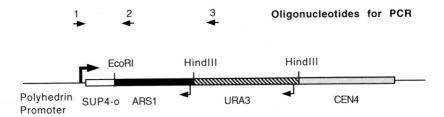

Oligonucleotides for PCR

Oligonucleotides for PCR

1. **Polyhedrin promoter**   5' CATGGAGATAATTAAAATGATAACCA 3'

2. **ARS**   5' CCCTAGTGCACTTACCCCACGTTCGG 3'

3. **URA3**   5' TTAGAGCTTCAATTTAATTATATCAG 3'

Figure 6. Location of primers for PCR screening of recombinant baculovirus DNA. A primer representing either the polyhedrin promoter, the ARS1, or the URA3 sequences is used for PCR analysis in conjunction with an oligonucleotide primer located within the inserted foreign cDNA.

expression are significantly lower. Thus, it is often wise to analyse extracts of infected cells harvested at different times post-infection, ranging from 24 hours to 72 hours.

Often, the degree of expression is such that sufficient protein can be obtained for further studies from relatively small scale production; for example, from a manageable number of tissue culture flasks or from a few one litre batches of cells growing in suspension. However, there will obviously be circumstances where large scale production is necessary. In the past, the growth and infection of the insect cells in large cultures has proved to be difficult. However, earlier problems have now largely been resolved and it is possible to use fermentors for the large scale production of baculovirus-mediated expression of proteins. These techniques are described elsewhere (3).

References

1. Sambrook, J., Fritsch, E. F., and Maniatis, T. (ed.) (1989). *Molecular cloning, a laboratory manual* (2nd edn). Cold Spring Harbor Laboratory Press, Cold Spring Harbor, NY.
2. Summers, M. D. and Smith, G. E. (1987). *A manual of methods for baculovirus vectors and insect cell culture procedures*. Texas Agricultural Experiment Station Bulletin No. 1555.

3. O'Reilly, D. R., Miller, L. K., and Luckow, V. A. (1992). *Baculovirus expression vectors: a laboratory manual*. W. H. and Company, NY.
4. Harlow, E. and Lane, D. (1988). *Antibodies: a laboratory manual*. Cold Spring Harbor Laboratory Press, Cold Spring Harbor, NY.
5. King, L. A. and Possee, R. D. (1992). *The baculovirus expression system: a laboratory guide*. Chapman and Hall, UK.
6. Arif, B. M. (1986). *Curr. Top. Microbiol Immunol.*, **131**, 21.
7. Luckow, V. A. and Summers, M. D. (1988). *Virology*, **167**, 56.
8. Blissard, G. W. and Rohrmann, G. F. (1990). *Annu. Rev Entomol.*, **35**, 127.
9. Maeda, S. (1989). *Annu. Rev. Entomol.*, **34**, 351.
10. Friesen, P. D. and Miller, L. K. (1986). *Curr. Top. Microbiol. Immunol.*, **131**, 31.
11. Granados, R. R., Lawler, K. A., and Burand, J. P. (1981). *Intervirology*, **16**, 71.
12. Rohrmann, G. F. (1986). *J. Gen. Virol.*, **67**, 1499.
13. Summers, M. D. and Smith, G. E. (1978). *Virology*, **84**, 390.
14. Faulkner, P. and Carstens, E. B. (1986). *Curr. Top. Microbiol. Immunol.*, **131**, 1.
15. Luckow, V. A. and Summers, M. D. (1988). *Bio/Technology*, **6**, 47.
16. Marais, R. M., Hsuan, J. J., McGuigan, C., Wynne, J., and Treisman, R. (1992). *EMBO J.*, **11**, 97.
17. Ollo, R. and Maniatis, T. (1987). *Proc. Natl. Acad. Sci. USA*, **84**, 5700.
18. Patel, G., Nasmyth, K., and Jones, N. (1992). *Nucleic Acids Res.*, **20**, 97.
19. Lin, B. T., Gruenwald, S., Morla, A. O., Lee, W. H., and Wang, J. Y. (1991). *EMBO J.*, **10**, 857.
20. Roy, L. M., Swenson, K. I., Walker, D. H., Gabrielli, B. G., Li, R. S., Piwnica-Worms, H., and Maller, J. L. (1991). *J. Cell. Biol.*, **113**, 507.
21. Patel, G. and Jones, N. C. (1990). *Nucleic Acids Res.*, **18**, 2909.
22. Piwnica-Worms, H., Williams, N. G., Cheng, S. H., and Roberts, T. M. (1990). *J. Virol.*, **64**, 61.
23. Greenfield, C., Patel, G., Clark, S., Jones, N., and Waterfield, M. D. (1988). *EMBO J.*, **7**, 139.
24. Herrera, R., Lebwohl, D., Garcia, d. H. A., Kallen, R. G., and Rosen, O. M. (1988). *J. Biol. Chem.*, **263**, 5560.
25. Patel, G. and Stabel, S. (1989). *Cell Signal*, **1**, 227.
26. Hoss, A., Moarefi, I., Scheidtmann, K. H., Cisek, L. J., Corden, J. L., Dornreiter, I., Arthur, A. K., and Fanning, E. (1990). *J. Virol.*, **64**, 4799.
27. Hughes, K., Nikolakaki E., Plyte, S. E., Totty, N. F., and Woodgett, J. R. (1993). *EMBO J.*, **12**, 803.
28. Kato, J., Matsushime, H., Hiebert, S. W., Ewen, M. E., and Sherr, C. J. (1993). *Genes Dev.*, **7**, 331.
29. Thomsen, D. R., Post, L. E., and Elhammer, A. P. (1990). *J. Cell. Biochem.*, **43**, 67.
30. Lanford, R. E. (1988). *Virology*, **167**, 72.
31. Page, M. J., Hall, A., Rhodes, S., Skinner, R. H., Murphy, V., Sydenham, M., and Lowe, P. N. (1989). *J. Biol. Chem.*, **264**, 19147.
32. Domingo, D. L. and Trowbridge, I. S. (1988). *J. Biol. Chem.*, **263**, 13386.
33. Lanford, R. E., Luckow, V., Kennedy, R. C., Dreesman, G. R., Notvall, L., and Summers, M. D. (1989). *J. Virol.*, **63**, 1549.
34. Maeda, S., Kawai, T., Obinata, M., Fujiwara, H., Horiuchi, T., Saeki, Y., Sato, Y., and Furusawa, M. (1985). *Nature*, **315**, 592.

35. Smith, G. E., Ju, G., Ericson, B. L., Moschera, J., Lahm, H. W., Chizzonite, R., and Summers, M. D. (1985). *Proc. Natl. Acad. Sci. USA*, **82**, 8404.
36. Jarvis, D. L. and Summers, M. D. (1989). *Mol. Cell. Biol.*, **9**, 214.
37. Kuroda, K., Groner, A., Frese, K., Drenckhahn, D., Hauser, C., Rott, R., Doerfler, W., and Klenk, H. D. (1989). *J. Virol.*, **63**, 1677.
38. Possee, R. D. (1986). *Virus Res.*, **5**, 43.
39. Pen, J., Welling, G. W., and Welling, W. S. (1989). *Nucleic Acids Res.*, **17**, 451.
40. Capone, J. (1989). *Gene Anal. Tech.*, **6**, 62.
41. Vlak, J. M., Schouten, A., Usmany, M., Belsham, G. J., Klinge, R. E., Maule, A. J., Van, L. J., and Zuidema, D. (1990). *Virology*, **179**, 312.
42. O'Reilly, D. R., Passarelli, A. L., Goldman, I. F., and Miller, L. K. (1990). *J. Gen. Virol.*, **71**, 1029.
43. Richardson, C., Lalumiere, M., Banville, M., and Vialard, J. (1992). In *Baculovirus expression protocols* (ed. C. Richardson and J. Walker). Humana: Clifton, NJ.
44. Zuidema, D., Schouten, A., Usmany, M., Maule, A. J., Belsham, G. J., Roosien, J., Klinge, R. E., Van, L. J., and Vlak, J. M. (1990). *J. Gen. Virol.*, **71**, 2201.
45. Kitts, P. A., Ayres, M. D., and Possee, R. D. (1990). *Nucleic Acids Res.*, **18**, 5667
46. Kitts, P. A. and Possee, R. D. (1993). *BioTechniques*, **14**, 810.
47. Matsuura, Y., Possee, R. D., Overton, H. A., and Bishop, D. H. (1987). *J. Gen. Virol.*, **68**, 1233.
48. Wang, X. Z., Ooi, B. G., and Miller, L. K. (1991). *Gene*, **100**, 131.
49. Luckow, V. A. and Summers, M. D. (1989). *Virology*, **170**, 31.
50. Beames, B., Braunagel, S., Summers, M. D., and Lanford, R. E. (1991). *BioTechniques*, **11**, 378.
51. Newman, A. and Norman, C. (1991). *Cell*, **65**, 115.

A1

Addresses of suppliers

Advanced Protein Products, Unit 18H, Premier Partnership Estate, Leys Road, Brockmoor, Brierley Hill, West Midlands DY5 3UP, UK.

American Type Culture Collection, 12301 Parklawn Drive, Rockville, MD 20852, USA.

Amersham International plc, Lincoln Place, Green End, Aylesbury, Buckinghamshire HP20 2TP, UK; 2636 Sth Clearbrook Drive, Arlington Heights, IL 60005, USA.

Amicon, Amicon Division, W. R. Grace and Co., 72 Cherry Hill, Beverley, MA 01915, USA.

AMRAD, Kew, Victoria, Australia.

AMS Biotechnology (UK) Ltd, 5 Thorney Leys Park, Witney, Oxon OX8 3DN, UK.

Anderman and Co. Ltd, 145 London Road, Kingston-Upon-Thames, Surrey KT17 7NH, UK.

B. Braun Biotech, 999 Postal Road, Allentown, PA 18103, USA.

Bayer Diagnostics, Evans House, Hamilton Close, Basingstoke, Hampshire RG21 2YE, UK.

Beckman Ltd, Progress Road, Sands Industrial Estate, High Wycombe, Buckinghamshire HP12 4JL, UK; 1050 Page Mill Road, Palo Alto, CA 94304, USA.

Becton Dickinson UK Ltd, Between Towns Road, Cowley, Oxford OX4 3LY, UK; Labware, 2 Bridgewater Lane, Lincoln Park, NJ 07035, USA.

Bio-Rad Laboratories Ltd, Bio-Rad House, Maylands Avenue, Hemel Hempstead, Hertfordshire HP2 7TD, UK; Alfred Nobel Drive, Hercules, CA 94547, USA.

Boehringer-Mannheim UK, Bell Lane, Lewes, East Sussex BN7 1LG, UK; Biochemical Products, 9115 Hague Road, PO Box 50414, Indianapolis, IN 46250–0414, USA.

Branson Ultrasonics Corporation, 41 Eagle Road, Danbury, CT, USA.

British Biotechnology Products Ltd, 4–10 The Quadrant, Barton Lane, Abingdon, Oxon OX14 3YS, UK.

Collaborative Medical Products Inc, Becton Dickinson Hardware, Two Oak Park, Bedford, MA 01730, USA.

Corning Inc, Science Products, MP-21-5-8, Corning, NY 14831, USA.

Dako, 16 Manor Courtyard, Hughendon Avenue, High Wycombe, Buckinghamshire HP13 5RE, UK; 6392 Via Real, Carpinteria, CA 93013, USA.

Difco Laboratories Ltd, PO Box 13B, Central Avenue, West Molesley, Surrey KT8 0SE, UK; PO Box 1058, Detroit, MI 48232, USA.

Dupont/NEN Research Products, 549 Albany Street, Boston, MA 02118, USA.

Falcon, (distributed in USA by Becton Dickinson Labware).

Gibco Life Technologies, 8400 Helgerman Court, Gaithersburg, MD 20877, USA.

Heat Systems Inc, 1938 New Hwy, Farmingdale, NY, USA.

Hybaid, 111–113 Waldegrave Road, Teddington, Middlesex TW11 8LL, UK; National Labnet Corporation, PO Box 841, Woodbridge, NJ 07095, USA.

ICN Biomedicals Inc, 3300 Hyland Avenue, Costa Mesa, CA 92626, USA.

International Biotechnologies Inc, (IBI Limited, A Kodak Company), 36 Clifton Road, Cambridge CB1 4ZR, UK; PO Box 9558, 25 Science Park, New Haven, CT 06535, USA.

Invitrogen Corporation, (distributed in UK by British Biotechnology Products Ltd); 3985 B Sorrento Valley Building, San Diego, CA 92121, USA.

Life Technologies, PO Box 35, Trident House, Renfrew Road, Paisley PA3 4FF, UK.

Merck-BDH, Merck Ltd, Poole, Dorset BH15 1TD, UK.

Millipore Corporation, 80 Ashby Road, Bedford, MA 01730, USA.

New England Biolabs/C.P. Laboratories, PO Box 22, Bishop's Stortford, Hertfordshire CM23 3DH, UK; 32 Tozer Road, Beverley, MA 01915–5599, USA.

Novagen Ltd, (see AMS Biotechnology (UK) Ltd as UK distributor); 597 Science Drive, Madison, WI 53711, USA.

Nunc A/S, PO Box, Kamstrup, DK-4000, Roskilde, Denmark.

Perkin-Elmer Cetus, 761 Main Avenue, Norwalk, CT 06856, USA.

Pharmacia, Davy Avenue, Knowlhill, Milton Keynes MK5 8PH, UK; PO Box 1327, Piscataway, NY 08855–1327, USA.

Pharmacia LKB Biotech Ltd, 23 Grosvenor Road, St. Albans, Hertfordshire AL1 3AW, UK; 800 Centennial Avenue, PO Box 1327, Piscataway, NJ 08854–1327, USA.

Promega Ltd, Delta House, Enterprise Road, Chilworth Research Centre, Southampton SO1 7NS, UK; 2800 Woods Hollow Road, Madison, WI 53711–5399, USA.

QIAGEN Inc, (distributed in UK by Hybaid); 9259 Eton Avenue, Chatsworth, CA 91311, USA.

Sarstedt Ltd, 68 Boston Road, Beaumont Leys, Leicester LE4 1AW, UK.

Schleicher and Schuell, (distributed in UK by Anderman and Co. Ltd); 10 Optical Avenue, PO Box 2012, Keene, NH 03431, USA.

Sigma Chemical Company Ltd, Fancy Road, Poole, Dorset BH17 7TG, UK; PO Box 14508, St. Louis, MO 63178, USA.

Sorvall Du Pont Ltd, Wedgewood Way, Stevenage, Hertfordshire SG1 4QN, UK; PO Box 80024, Wilmington, DE 19880–0024, USA.

Sterogene Biochemicals, San Rafael, California, USA.

Stratagene Ltd, Cambridge Innovation Centre, Cambridge Science Park, Milton Road, Cambridge CB4 4GF, UK; 11099 North Torrey Pines Road, La Jolla, CA 92037, USA.

Techne Ltd, Duxford, Cambridge CB2 2PZ, UK.

Vector Laboratories Inc, 30 Ingold Road, Burlingame, CA 94010, USA.

Index

Companion volumes available from IRL Press at Oxford University Press

Editors: David M. Glover and B. David Hames

DNA Cloning 1: Core Techniques

DNA Cloning 2: Expression Systems

DNA Cloning 3: Complex Genomes

1. Cosmid clones and their application to genome studies
   Alasdair C. Ivens and Peter F. R. Little
2. Chromosome-specific gridded cosmid libraries: construction, handling and use in parallel and integrated mapping
   Dean Nizetic and Hans Lehrach
3. Library construction in P1 phage vectors
   Nat Sternberg
4. Cloning into yeast artificial chromosomes
   Rakesh Anand
5. Amplification of DNA microdissected from mitotic and polytene chromosomes
   Robert D. C. Saunders
6. Databases, computer networks and molecular biology
   Rainer Fuchs and Graham N. Cameron
7. Long range restriction mapping
   Wendy Bickmore
8. Genetic mapping with microsatellites
   Isam S. Naom, Christopher G. Mathew, and Margaret-Mary Town

DNA Cloning 4: Mammalian Systems

1. High efficiency gene transfer into mammalian cells
   Vanessa Chisholm
2. Construction and characterization of vaccinia virus recombinants
   Mike Mackett
3. Vectors based on gene amplification for the expression of cloned genes in mammalian cells
   Christopher R. Bebbington
4. Retroviral vectors
   Anthony M. C. Brown and Joseph P. Dougherty
5. Introduction of cloned DNA into embryonic stem cells
   Amy R. Mohn and Beverly H. Koller
6. Production of transgenic rodents by the microinjection of cloned DNA into fertilized one-cell eggs
   Sarah Jane Waller, Mei-Yin Ho, and David Murphy
7. Genomic and expression analysis of transgenic animals
   Sarah Jane Waller, Judith McNiff Funkhouser, Kum-Fai Chooi, and David Murphy
8. Expression of cloned DNA using a defective Herpes simplex virus (HSV-1) vector system
   Filip Lim, Philip Starr, Song Song, Dean Hartley, Phung Lang, and Alfred I. Geller
9. Adenovirus vectors
   Robert D. Gerard and Robert S. Meidell